AF597445

INTRODUCTION TO FLUID FLOW AND THE TRANSFER OF HEAT AND MASS

PRENTICE-HALL INTERNATIONAL SERIES
IN THE PHYSICAL AND CHEMICAL ENGINEERING SCIENCES

NEAL R. AMUNDSON, EDITOR, *University of Houston*

AMUNDSON *Mathematical Methods in Chemical Engineering: Matrices and Their Applications*
BALZHIZER, SAMUELS, AND ELLIASSEN *Chemical Engineering Thermodynamics*
BRIAN *Staged Cascades in Chemical Processing*
BUTT *Reaction Kinetics and Reactor Design*
DENN *Process Fluid Mechanics*
FOGLER *The Elements of Chemical Kinetics and Reactor Calculations: A Self-Paced Approach*
FOGLER AND BROM *Elements of Chemical Reaction Engineering*
HIMMELBLAU *Basic Principles and Calculations in Chemical Engineering, 4th edition*
HINES AND MADDOX *Mass Transfer: Fundamentals and Applications*
HOLLAND *Fundamentals and Modeling of Separation Processes: Absorption, Distillation, Evaporation, and Extraction*
HOLLAND AND ANTHONY *Fundamentals of Chemical Reaction Engineering*
KUBICEK AND HLAVACEK *Numerical Solution of Nonlinear Boundary Value Problems with Applications*
KYLE *Chemical and Process Thermodynamics*
LEVICH *Physiochemical Hydrodynamics*
MODELL AND REID *Thermodynamics and its Applications, 2nd edition*
MYERS AND SEIDER *Introduction to Chemical Engineering and Computer Calculations*
OLSON AND SHELSTAD *Introduction to Fluid Flow and the Transfer of Heat and Mass*
PRAUSNITZ, LICHTENTHALER, AND DE AZEVEDO *Thermodynamics of Fluid-Phase Equilibria, 2nd edition*
PRAUSNITZ ET AL *Computer Calculations for Multicomponent Vapor-Liquid and Liquid-Liquid Equilibria*
RAMKNSHNA AND AMUNDSON *Linear Operation Methods in Chemical Engineering with Applications for Transport and Chemical Reaction Systems*
RHEE ET AL *First-Order Partial Differential Equations: Theory and Applications of Single Equations*
RUDD ET AL *Process Synthesis*
SCHULTZ *Diffraction for Materials Scientists*
STEPHANOPOULOS *Chemical Process Control: An Introduction to Theory and Practice*
VILLADSEN AND MICHELSEN *Solution of Differential Equation Models by Polynomial Approximation*

INTRODUCTION TO FLUID FLOW AND THE TRANSFER OF HEAT AND MASS

A. T. Olson
K. A. Shelstad
University of Western Ontario

PRENTICE-HALL, INC., Englewood Cliffs, New Jersey 07632

Library of Congress Cataloging-in-Publication Data

OLSON, A. T.
Introduction to fluid flow and the transfer of heat and mass.

Bibliography: p.
Includes index.
1. Fluid mechanics. 2. Heat—Transmission.
3. Mass transfer. I. Shelstad, K. A., (date).
II. Title.
QC145.2.048 1987 620.1'06 86-17068
ISBN 0-13-483892-0

Editorial/production supervision and
interior design: *Joan McCulley*
Manufacturing buyer: *Rhett Conklin*

Printed in the United States of America

10 9 8 7 6 5 4 3 2 1

ISBN 0-13-483892-0 025

PRENTICE-HALL INTERNATIONAL (UK) LIMITED, *London*
PRENTICE-HALL OF AUSTRALIA PTY. LIMITED, *Sydney*
PRENTICE-HALL CANADA INC., *Toronto*
PRENTICE-HALL HISPANOAMERICANA, S.A., *Mexico*
PRENTICE-HALL OF INDIA PRIVATE LIMITED, *New Delhi*
PRENTICE-HALL OF JAPAN, INC., *Tokyo*
PRENTICE-HALL OF SOUTHEAST ASIA PTE. LTD., *Singapore*
EDITORA PRENTICE-HALL DO BRASIL, LTDA., *Rio de Janeiro*

Contents

	Preface	ix
1	Introduction	1
1.1	Systems of Measurement	1
1.2	Molar Units	4
1.3	Ideal Gas Equation	6
1.4	Transport Processes	7
1.5	Ratio of Diffusivities	13
	References	13
	Problems	14
2	Introduction to Fluid Mechanics	17
2.1	Fluid Properties	17
2.2	Behavior of Fluids at Rest	18
2.3	Description of Flow	21
2.4	Equation of Continuity (Conservation of Mass)	24
2.5	Relationship between Pressure and Velocity	25
2.6	Linear Momentum Equation	27
2.7	Flow Measurement	29
	References	39
	Problems	39
3	Applications in One-Dimensional Fluid Flow	44
3.1	Effect of Friction on the Pressure–Velocity Relation	44
3.2	Steady-Flow Energy Equation	44

3.3 Kinetic Energy Correction Factor α 46
3.4 Dimensional Analysis Applied to Pipe Friction 47
3.5 Problem Types Using the Friction Factor 52
3.6 Laminar Flow in Pipes 55
3.7 Minor Losses in Pipe Flow 58
3.8 Some Pipe-Flow Problems and Solutions 62
3.9 Drag and Lift 66
3.10 Preview of Two-Dimensional Fluid Flow 69
Problems 72

4 Fluid Flow in Boundary Layers 76

4.1 Introduction to Boundary Layer Flow 76
4.2 Displacement Thickness δ_1 (Mass-Flow Deficit) 77
4.3 Momentum Thickness δ_2 (Momentum Deficit) 78
4.4 Kinetic Energy Thickness δ_3 (Kinetic Energy Deficit) 78
4.5 Notes Regarding the Boundary Layer Thickness and Integration Limits 79
4.6 Summary of Boundary Layer Thickness for Incompressible Flow 79
4.7 Boundary Layer Equations 82
4.8 Mass Conservation Equation 82
4.9 Momentum Equation 83
4.10 Discussion of Boundary Conditions 84
4.11 Energy Equation 85
4.12 Summary of Boundary Layer Equations for Incompressible Flow 88
4.13 Von Kármán Momentum Integral 88
4.14 Laminar Boundary Layer Growth on a Flat Plate 91
4.15 Choice of a Suitable Velocity Profile 93
4.16 Turbulent Boundary Layer Shape and Wall Shear Stress 95
4.17 Turbulent Boundary Layer Growth on a Flat Plate 98
Problems 101

5 Steady-State Conduction of Heat 103

5.1 Fourier's Heat Conduction Equation 103
5.2 Heat Conduction through a Plane Wall 104
5.3 Heat Conduction through a Composite Wall 105
5.4 Surface Coefficient of Heat Transfer 106
5.5 Heat Conduction through a Thick-Walled Tube 108

5.6	Heat Conduction through a Composite Cylinder	110
5.7	Heat Conduction through a Thick-Walled Sphere	111
5.8	Heat Loss from Insulated Pipes	112
5.9	Conduction in Walls with Heat Generation	113
5.10	Convection at Walls with Heat Generation	117
5.11	Conduction in Cylinders with Heat Generation	118
5.12	Heat Transfer from Fins	120
5.13	Rate of Heat Transfer from a Fin	124
5.14	Fin Efficiency	126
	References	129
	Problems	129

6 Convective Heat Transfer 133

6.1	Introduction	133
6.2	First Law Method for Finding *h*	134
6.3	Temperature Gradient at Solid–Fluid Interface	135
6.4	Determination of *h* Using Boundary Layer Analysis	137
6.5	Suggested Perspective	143
6.6	Forced Convection Correlation	144
6.7	Natural Convection Correlation	146
6.8	Empirical Correlations for Convective Heat Transfer	146
	References	153
	Problems	154

7 Heat Exchangers 157

7.1	Double-Pipe Heat Exchanger	155
7.2	Mean Temperature Difference for a Double-Pipe Exchanger	160
7.3	Multipass Exchangers	162
7.4	Prediction of the Overall Coefficient *U*	166
7.5	Fouling of Heat-Transfer Surfaces	167
	References	167
	Problems	167

8 Radiant Heat Transfer 170

8.1	Introduction	170
8.2	Some Definitions and Properties	171
8.3	Black-Body Radiation	172

8.4 Kirchhoff's Law 173
8.5 Monochromatic Emissive Power and Spectral Energy Distribution 174
8.6 Radiant Heat Transfer between Black Bodies 176
8.7 Radiation Intensity 177
8.8 Geometric Shape Factor, F_{1-2} 180
8.9 Shape Factor Relationships 181
8.10 Sample Shape Factor Calculations 183
8.11 Geometric Flux Algebra 185
8.12 Special Reciprocity Relation 186
8.13 Radiant Heat Exchange between Black Bodies 189
8.14 Real Surfaces and Gray Surfaces 190
8.15 Radiant Heat Exchange between Gray Bodies 191
8.16 Electrical Analog for Radiating Systems 192
References 200
Problems 200

9 Steady-State Diffusion of Mass 205

9.1 Fick's Law of Diffusion 205
9.2 Diffusion Coefficient 206
9.3 Stefan's Experiment 208
9.4 Steady-State Diffusion through a Stagnant Gas 210
9.5 Equimolar Counterdiffusion 213
9.6 Diffusion into an Infinite Stagnant Medium 214
References 220
Problems 220

10 Convective Mass Transfer 223

10.1 Mass Transfer within a Single Fluid Phase 224
10.2 Mass-Transfer Coefficient at Surfaces of Simple Geometry 228
10.3 Mass Transfer under Conditions of Natural Convection 234
10.4 Mass Transfer between Two Fluids 240
10.5 Estimate of Tower Size Required to Remove SO_2 from Air 250
References 252
Problems 253

APPENDIX 254

INDEX 261

Preface

This book presents an elementary treatment of the principles of fluid mechanics and of heat and mass transfer. As a text it contains more than enough material for a one semester course at the sophomore level, thus permitting course instructors some choice in the subject matter to be covered in their individual courses. A first course in calculus as background is essential for the proper understanding of the material.

The book begins with a discussion of units and the different systems of measurement. First, the point is made that most engineers and practicing technologists must be familiar with more than one system of measurement. The three processes of momentum, heat, and mass transfer are then introduced along with the defining equations for one-dimensional transport.

The presentation on fluid mechanics includes the basic relationships involving continuity, energy, and momentum; their application to metering devices, such as the Pitot tube and the orifice meter; and a discussion of systems, such as flow through pipes, where friction cannot be ignored. The concept of the boundary layer is introduced in Chapter 4.

The material on heat transfer covers one-dimensional conduction in solids of simple shape, conduction and convection from fins, heat exchangers, and radiation heat transfer. Convective heat transfer is discussed in Chapter 6. Instructors might choose to omit from an introductory course some of the material in Chapters 4 and 6 on the boundary layer.

The final two chapters discuss mass transfer by molecular diffusion and by convection.

Problems at the end of each chapter illustrate the application of ideas discussed in the text. An effort has been made to present realistic problems in order for the student to obtain some insight into the application of the text material to problems in industry.

Finally, the authors wish to acknowledge the contributions of many persons to the final manuscript. For typing and seeing the manuscript through several revisions, we extend our appreciation to Gusta Johnson, Clare Whittaker, Cindy Reid, Janet Greenfield, and Connie Sprague. While the manuscript was being developed, its text was used for several years as the basis of a course given to all engineering students in their second year of the program. From discussions with these students, many clarifications to the text and errors in the solutions to the problems were discovered and corrected. Any errors that remain, and hopefully there are none, are the responsibility of the authors. Finally, a word from one student who commented at the end of her final examination, "And in the beginning there was much momentum, heat, and mass transfer."

A. T. Olson
K. A. Shelstad

Introduction 1

In this chapter our main purpose is to introduce the concept of rate and to relate it to the processes of momentum, heat, and mass transfer. However, before doing this we want to ensure that the measurement systems to be used in the calculations are clearly understood. A number of these systems of measurement have been developed and used in the past. An attempt to supercede these by a standard system known as the Système Internationale d'Unités (SI system) is underway at the present time. We will emphasize the use of the SI system in the material of this book. Nevertheless, we believe it is important that engineers retain a working knowledge of the older systems and be able to convert from one system to the other. Industrial plants built in past years to English engineering specifications will continue to need operating and management personnel who can work with this system for years to come. Also, the older literature of engineering that did not use the SI system contains much information that can be used in the designs of the future. Our first concern, then, is to be certain that we understand these different systems of measurement with the corresponding dimensions and units that are used by engineers.

1.1 Systems of Measurement

The various systems of measurement that are pertinent to the subject of this book derive from the choice made for the basic dimensions. The basic dimensions in mechanical systems are generally considered to be force (F), mass (M), length (L), and time (t). Newton's second law of motion provides a relation between these four dimensions.

$$\text{Force} \propto \text{Mass} \times \text{Acceleration}$$

We can express this statement in the form of an equality by insertion of a proportionality constant.

$$\text{Force} = \text{constant} \times \text{mass} \times \text{acceleration} \tag{1.1}$$

To be dimensionally sound, this equation must have the same dimensions on each side of the equal sign; that is, dimensionally,

$$F = \text{constant} \times M \times Lt^{-2} \tag{1.2}$$

The various systems of measurement in common use derive from three basic decisions, a decision regarding the constant in Newton's second law, a choice of the physical quantities to be considered as basic, and a choice of unit sizes for measuring the basic quantities.

1.1.1 English Engineering System. In the English engineering system, we define independently the four basic quantities of mass, force, length, and time.

Quantity	Dimensions	Units
Mass	M	lb_m
Force	F	lb_f
Length	L	ft
Time	t	sec

The proportionality constant in Newton's second law is written as $1/g_c$, where g_c is called the *universal gravitational constant*. This constant has a magnitude and dimensions such that when Newton's second law is written as

$$F = \frac{1}{g_c} ma$$

the calculated force will correspond correctly to the size of the mass and acceleration, and the units and dimensions will be the same on both sides of the equation; that is,

$$g_c = \frac{ma}{F} = 32.17 \frac{lb_m\text{-ft}}{lb_f\text{-sec}^2}$$

1.1.2 Systems That Eliminate g_c. Units and dimensions can be defined so that Newton's second law is simply $F = ma$. Such systems have evolved and are in common use. Three of the four basic quantities are arbitrarily defined, and the fourth is derived in terms of the other three with the aid of Newton's second law such that the proportionality constant is unity and dimensionless. Systems in which mass, length, and time are defined are called *absolute*. Systems in which force, length, and time are defined are called *gravitational*.

1.1.3 Système International d'Unités (SI).

SI is an absolute system in which mass, length, and time are defined. The concept of force is derived from Newton's second law such that the proportionality constant is unity and dimensionless. Newton's second law is written in the form $F = ma$.

Quantity		Dimensions	Units
Defined:	mass	M	kg
	length	L	m
	time	t	s
Derived:	force	$(F) = MLt^{-2}$	$(\text{N}) = \dfrac{\text{kg-m}}{\text{s}^2}$

A newton (N) is defined as the force required to give a 1-kilogram (kg) mass an acceleration of 1 meter per second^2 (m/s^2).

1.1.4 British Gravitational System.

In the British gravitational system, force, length, and time are defined as basic, and the concept of mass is derived such that $F = ma$.

Quantity		Dimensions	Units
Defined:	force	F	lb
	length	L	ft
	time	t	sec
Derived:	mass	$(M) = FL^{-1}t^2$	$(\text{slug}) = \dfrac{\text{lb-sec}^2}{\text{ft}}$

The unit of mass is 1 slug, equivalent to 32.17 lb mass. The unit slug is equivalent to the compound unit lb-sec^2/ft.

1.1.5 Other Systems of Units.

The English engineering system, the SI, and the British gravitational system are the systems commonly encountered in engineering calculations. However, physical data derived from the technical literature may also be expressed in the absolute metric system (cgs) or the English absolute system. In the cgs system, mass (grams), length (cm), and time (sec) are defined, and the concept of force (dyne) is derived. In the English absolute system, mass (lb), length (ft), and time (sec) are defined, and the unit of force (poundal) is derived. One poundal is equal to (1/32.17)lb_f.

1.1.6 Thermal Units. When thermal effects are present, temperature (T) must be included among the dimensions describing the system. Heat has dimensions of energy ($F \times L$), although it is permissible to include heat (H) as an additional dimension. If the dimension H is used, a dimensional constant called the *mechanical equivalent of heat* must be included in the analysis to correlate heat units with those of energy (work).

In the SI system, the amount of heat is expressed as joules (J), which is the same as work, where 1 joule = 1 newton-meter.

In the absolute metric system (cgs), the unit of heat is the gram calorie (cal), defined as the amount of heat required to raise the temperature of 1 gram of water by 1°C (1 cal = 4.1868 J).

In the various English systems, the British thermal unit (Btu) is the amount of heat required to raise 1 pound of water from 60° to 61°F.

A summary of the units of a number of quantities in the various systems of measurement is given in Table 1.1.

1.1.7 Conversion of Units. Simple conversions from one set of units to another can be carried out using conversion factors such as the following:

Mass	$1\ \text{lb}_m = \left(\dfrac{1}{32.17}\right)$ slug = 453.6 g = 0.4536 kg
Length	1 ft = 12 in. × 2.54 (cm/in.) = 30.48 cm = 0.3048 m
Time	1 hr = 3600 sec
Force	$1\ \text{lb}_f = 32.17$ poundals = 4.44×10^5 dyne = 4.44 newtons
Temperature	1 C° = 1 K° = 1.8 F°
Pressure	1 atm = 14.7 psi = 1.013×10^5 Pa = 760 mm Hg

1.2 Molar Units

In certain types of diffusion problems or where chemical reactions are involved, it is usually convenient to work in terms of molar units. The kilogram mole (kg mol) is the amount of substance whose mass in kilograms is numerically equal to its molecular weight. The gram mole (g mol) and pound mole (lb mol) are similarly defined. The molecular weights of some common substances are listed in Table A.5 in the Appendix.

Example 1.2.1

The density of benzene at 15°C is 880 kg/m³. What is the molar density of benzene? Express the density in the cgs and English engineering systems of units. From Table A.5, the molecular weight of benzene is 78.11. The molar density of benzene is

$$C = \frac{880}{78.11} = 11.3\ \text{kg mol/m}^3$$

Table 1.1

Units of Quantities in Different Systems of Measurement

Quantity	English Engineering	Absolute			English Gravitational
		SI	English	cgs	
Mass	lb_m	kg	lb	g	slug
Length	ft	m	ft	cm	ft
Time	sec	s	sec	sec	sec
Force	lb_f	newton (N)	poundal	dyne	lb
g_c	$32.17 \frac{lb_m\text{-ft}}{lb_f\text{-sec}^2}$	1	1	1	1
Energy or work	ft-lb_f	joule (J)	ft-poundal	erg	ft-lb
Pressure	lb_f/ft^2	newton/m^2 (pascal, Pa)	poundal/ft^2	dyne/cm^2	lb/ft^2
Power	ft-lb_f/sec	watt	ft-poundal/sec	erg/sec	ft-lb/sec
Temperature	°F	K	°F	°C	°F
Heat	Btu	joule (J)	Btu	cal	Btu
Specific heat	Btu/lb_m-°F	joule/kg-K	Btu/lb-°F	cal/g-°C	Btu/slug-°F
Mechanical equivalent of heat	778 ft-lb_f/Btu	1	2.50×10^4 ft-poundals/lb-°F	4.18×10^7 erg/g-°C	778 ft-lb/Btu

In the cgs system,

$$\rho = 880\,\frac{\text{kg}}{\text{m}^3} \times \frac{1000\text{ g}}{\text{kg}} \times \left(\frac{\text{m}}{100\text{ cm}}\right)^3 = 0.880\text{ g/cm}^3$$

$$C = \frac{0.880}{78.11} = 0.0113\,\frac{\text{g mol}}{\text{cm}^3}$$

In the English engineering system,

$$\rho = 880\,\frac{\text{kg}}{\text{m}^3} \times \frac{\text{lb}_\text{m}}{0.4536\text{ kg}} \times \left(\frac{12 \times 2.54}{100}\,\frac{\text{m}}{\text{ft}}\right)^3 = 54.9\,\frac{\text{lb}_\text{m}}{\text{ft}^3}$$

$$C = \frac{54.9}{78.11} = 0.703\,\frac{\text{lb mol}}{\text{ft}^3}$$

Example 1.2.2

Express a pressure of 14.7 psi in the cgs and SI systems of units.

$$\text{cgs } P = 14.7\,\frac{\text{lb}_\text{f}}{\text{in.}^2} \times \frac{4.44 \times 10^5\text{ dynes}}{\text{lb}_\text{f}} \times \left(\frac{\text{in.}}{2.54\text{ cm}}\right)^2 = 10.1 \times 10^5\text{ dynes/cm}^2$$

$$\text{SI } P = 10.1 \times 10^5\,\frac{\text{dynes}}{\text{cm}^2} \times \frac{\text{N}}{10^5\text{ dynes}} \times \left(\frac{100\text{ cm}}{\text{m}}\right)^2 = 10.1 \times 10^4\text{ N/m}^2 = 10.1 \times 10^4\text{ Pa}$$

1.3 Ideal Gas Equation

An ideal gas has the following equation of state:

$$PV = n\,RT = \frac{m}{M}\,RT \tag{1.3}$$

where n is the number of moles of gas and m the corresponding mass of gas of molecular weight M contained in the volume V at the absolute pressure P and absolute temperature T. R is a constant whose numerical value depends only on the system of units used. Some commonly used values for R are listed in Table 1.2.

TABLE 1.2
VALUES OF THE GAS CONSTANT R

R	Units of P	Units of V	Units of T	Units of n
82.05	atm	cm^3	°K	g mol
1545	$\text{lb}_\text{f}/\text{ft}^2$	ft^3	°R	lb mol
0.7302	atm	ft^3	°R	lb mol
8314	Pa	m^3	K	kg mol

No real gas follows equation (1.3) exactly. However, all gases approach ideal gas behavior as the pressure is decreased. For the purpose of engineering calculations, use of (1.3) usually gives sufficient accuracy except at high pressures and low temperatures where deviations from ideal behavior become significant.

Example 1.3.1

A steel vessel of volume 2.0 liters contains air at an absolute pressure of 1.7×10^4 kPa and a temperature of 20°C. What is the mass of the air in the vessel?

$$R = 8314 \frac{\text{Pa m}^3}{\text{kg mol K}}$$

If deviations from ideal behavior are neglected, the number of moles of air in the vessel is:

$$n = \frac{PV}{RT} = \frac{1.7 \times 10^4 \times 10^3 \times 2.0 \times 10^{-3}}{8314\ (273 + 20)} = 1.4 \times 10^{-2} \text{ kg mol}$$

From Table A.5, the molecular weight of air is 28.9. Thus, the mass of air in the vessel is

$$\text{mass} = 1.4 \times 10^{-2} \times 28.9 = 0.40 \text{ kg}$$

1.4 Transport Processes

A system not in equilibrium has a tendency to attain that condition. In doing so, one or more quantities of the system are transported from one part of the system to another. For instance, a volume of liquid with a distribution of temperature is not in an equilibrium condition. In approaching that condition, heat is transported spontaneously from regions of high temperature to regions of lower temperature until eventually the temperature is uniform throughout the entire volume.

In the study of transport processes, it is the *rate of transport*[1] of a particular quantity that is of interest. This rate can be characterized as a driving force or potential difference divided by a resistance.

$$\text{Rate of transfer} = \frac{\text{driving force}}{\text{resistance}} \tag{1.4}$$

Expression (1.4) is similar to Ohm's law in physics, where the flow of electrical charge is proportional to the voltage difference and inversely proportional to

[1] In certain instances, it is convenient to refer to the flux rather than to the rate of transport. The *flux* is the rate of transport of a particular quantity per unit of area, where the area is taken normal to the direction of transfer.

the electrical resistance. When applied to the transport of momentum, heat, or mass, we will have to define each term in (1.4) in accordance with an understanding of the physics of each particular transport process.

1.4.1 Newton's Equation of Viscosity. When we stir a volume of liquid, for instance, by means of a paddle, the liquid is set in motion in such a way that certain parts are moving more quickly than other parts. If we stop stirring, the motion is soon damped out and the liquid comes to rest. Such an everyday experience illustrates that there are forces acting within a body of liquid that dissipate the energy associated with the liquid motion. These frictionlike forces are a manifestation of a property called *viscosity*, which is a characteristic quality of all liquids and gases.

We can obtain a more precise definition of what we mean by the viscosity of a fluid through an interpretation of the following imaginary experiment. Suppose a fluid is placed between two large parallel plates of area A separated by a distance h, as illustrated in Figure 1.1. The upper plate is held stationary while the lower plate moves with the steady velocity V in the x direction. If there is *no slip* between the fluid and the plates, the fluid in immediate contact with a plate has the same velocity as the plate. This condition, together with the *internal friction* or viscous forces acting between layers of fluid, results in a gradient of velocity being established through the layer of fluid. The oblique line in Figure 1.1 shows this velocity gradient to be a constant. The fluid is said to be in a state of *shear*, the shearing forces F indicated in Figure 1.1 being those required to maintain the velocity gradient.

Numerous investigators have shown that for many fluids there is a direct proportionality between the shear stress and the velocity gradient. The shear stress τ is defined as the ratio F/A, where F is taken as the shear force acting in the positive x direction. Mathematically, the relationship between shear stress and velocity gradient is

$$\tau \propto \frac{\Delta V}{h} \tag{1.5}$$

If a constant of proportionality is included in (1.5) and the velocity gradient

Figure 1.1 Viscous flow between two large, parallel plates

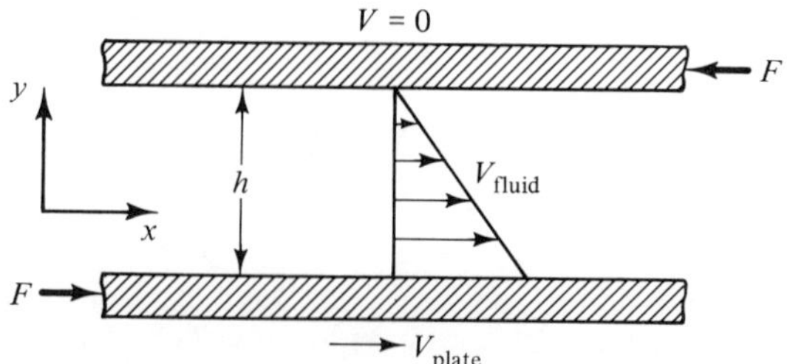

written as a derivative, the following equation is obtained:

$$\tau = -\mu \frac{dV}{dy} \tag{1.6}$$

The negative sign in equation (1.6) is required to conform to the situation illustrated in Figure 1.1, where the velocity is seen to decrease in the positive y direction. The proportionality factor μ is called the *absolute* or *dynamic viscosity* of the fluid. Fluids for which μ is independent of the rate of shear give the steady-state linear velocity profile illustrated in Figure 1.1 and are commonly referred to as Newtonian-type fluids.

It is of interest to look at the dimensions of τ, making use of Newton's second law of motion.

$$\tau = \frac{F}{A} \quad \text{and} \quad F = ma = \frac{d(mV)}{dt}$$

Shear stress is thus seen to have the same dimensions as momentum per unit time per unit area, which is momentum flux. Physically, this suggests that the action of the shear force F is to generate flow of momentum in the y direction. We can therefore say that Newton's equation of viscosity states that momentum flux is proportional to the negative velocity gradient in the fluid.

The dimensions of μ are seen from (1.6) to be FtL^{-2}. Table 1.3 lists the units of μ in three widely used systems of units. In the cgs system, 1 g/cm-sec is called a *poise*. Because this is a large unit for many fluids, viscosity data are often presented in centipoise (cp). In the English engineering system, the two sets of units for μ in Table 1.3 are related by the conversion factor g_c.

$$\mu \frac{\text{lb}_\text{f}\text{-sec}}{\text{ft}^2} \equiv \mu \frac{\text{lb}_\text{m}}{\text{ft-sec}} \times \frac{1}{g_c} \frac{\text{lb}_\text{f}\text{-sec}^2}{\text{lb}_\text{m}\text{-ft}}$$

For certain calculations, it is convenient to use the kinematic viscosity rather than the dynamic viscosity. This is defined as follows:

$$\nu = \frac{\mu}{\rho}$$

where ρ is the mass density. The dimensions of ν are L^2t^{-1}, so the dimensions of μ must be $ML^{-1}t^{-1}$.

TABLE 1.3
UNITS OF DYNAMIC VISCOSITY μ

SI	kg/m-s
cgs	dyne-sec/cm^2 or g/cm-sec
English engineering	lb_f-sec/ft^2 or lb_m/ft-sec

Numerical values for the viscosity of water and air are given in Tables A.1 and A.2. The values for air are seen to be smaller than those for water. It should also be noted that the viscosity of air increases with temperature, while that of water decreases with temperature.

Equation (1.6) is a valid function for all gases and many liquids, which are therefore referred to as Newtonian-type fluids. However, many industrially important liquids, such as paints, clay suspensions, and molten plastics, do not follow equation (1.6). The flow characteristics of such non-Newtonian liquids are beyond the scope of the present discussion.

Example 1.4.1

The viscosity of water at 21°C is 9.8×10^{-4} kg/m-s. Express this in terms of the other units listed in Table 1.3.

English engineering system:

$$\mu = 9.8 \times 10^{-4} \frac{\text{kg}}{\text{m-s}} \times \frac{\text{lb}_\text{m}}{0.4536 \text{ kg}} \times \frac{12 \times 2.54}{100} \frac{\text{m}}{\text{ft}} = 6.6 \times 10^{-4} \frac{\text{lb}_\text{m}}{\text{ft-sec}}$$

$$= 6.6 \times 10^{-4} \frac{\text{lb}_\text{m}}{\text{ft-sec}} \times \frac{1}{32.17} \frac{\text{lb}_\text{f}\text{-sec}^2}{\text{lb}_\text{m}\text{-ft}} = 2.1 \times 10^{-5} \frac{\text{lb}_\text{f}\text{-sec}}{\text{ft}^2}$$

cgs system:

$$\mu = 9.8 \times 10^{-4} \frac{\text{kg}}{\text{m-s}} \times \frac{1000 \text{ g}}{\text{kg}} \times \frac{\text{m}}{100 \text{ cm}} = 9.8 \times 10^{-3} \frac{\text{g}}{\text{cm-sec}}$$

$$= 9.8 \times 10^{-3} \text{ poise} = 0.98 \text{ cp} = 9.8 \times 10^{-3} \frac{\text{dyne-sec}}{\text{cm}^2}$$

1.4.2 Fourier's Equation of Heat Conduction. Let us consider a solid, homogeneous slab of area A with one face held at temperature T_1 and the other at temperature T_2 (Figure 1.2). With $T_1 > T_2$, experience teaches us that heat will flow spontaneously from the region of high temperature to the region of lower temperature. We also learn from experience that, when conditions are steady, the rate at which heat flows through the slab per unit of slab area, called the *heat flux*, is directly proportional to the temperature gradient. That is,

$$\frac{q}{A} \propto \frac{T_1 - T_2}{h} \tag{1.7}$$

If we put in a constant of proportionality and write the temperature gradient as a derivative, we obtain

$$\frac{q}{A} = -k \frac{dT}{dx} \tag{1.8}$$

where k is called the *thermal conductivity*. The negative sign indicates that

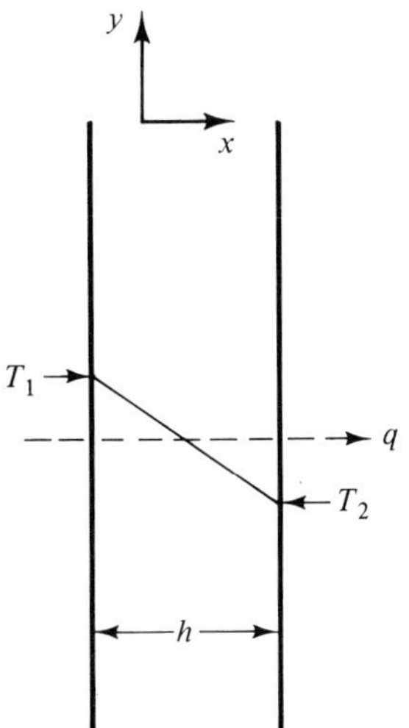

Figure 1.2 Conduction of heat through a solid slab

the temperature gradient is negative if x is taken as positive in the direction of heat flow. The oblique line in Figure 1.2 shows the temperature gradient to be a constant, which is true only if k is a constant. Equation (1.8) is known as Fourier's equation of heat conduction. We see it is similar mathematically to Newton's equation of viscosity, equation (1.6). The thermal conductivity k is a property of the material through which the heat is conducted. Some values are listed in Tables A.1, A.2, and A.3 for water, air, and a few solids. It should also be noted that the area A in equation (1.8) is the area at right angles to the direction in which the heat is flowing.

A quantity known as the *thermal diffusivity* is sometimes used in the analysis of heat-transfer problems. It is defined as

$$\alpha = \frac{k}{\rho c}$$

where ρ is the mass density and c the specific heat. Note that the dimensions of α are the same as for kinematic viscosity, that is, L^2t^{-1}.

Example 1.4.2

The thermal conductivity of aluminum metal at 100°C is given in the Appendix as 206 W/m°C. Express this in the English engineering and cgs systems of measurement.

English Engineering System

$$\text{k} = 206\,\frac{\text{W}}{\text{m °C}} \times \frac{3.413\text{ Btu}}{\text{W hr}} \times \frac{12 \times 2.54}{100}\frac{\text{m}}{\text{ft}} \times \frac{\text{°C}}{1.8\text{°F}} = 119\,\frac{\text{Btu}}{\text{hr-ft °F}}$$

cgs system

$$k = 206 \frac{\text{W}}{\text{m °C}} \times \frac{\text{cal}}{4.1868 \text{ W sec}} \times \frac{\text{m}}{100 \text{ cm}} = 0.492 \frac{\text{cal}}{\text{sec-cm °C}}$$

Note: Conversion factors such as 3.413 Btu/hr-W and 4.1868 joules/cal are to be found in various handbooks (1,2,3).

1.4.3 Fick's Equation of Diffusion.. Suppose we imagine a fluid confined between two parallel plates with the substance i of the upper plate being slowly dissolved in the fluid and then precipitated onto the lower plate (Figure 1.3). Our experience teaches us two things, first, that the substance i moves spontaneously from a region of high to one of lower concentration, and secondly, that the flux of i when conditions are steady is proportional to the concentration gradient. Stated mathematically for $C_{i1} > C_{i2}$:

$$\frac{n_i}{A} \propto \frac{C_{i1} - C_{i2}}{h} \tag{1.9}$$

where n_i/A is the molar flux of i and C_i is the molar concentration of i. If we put in a constant of proportionality and express the concentration gradient as a derivative we obtain the following expression:

$$J_i \equiv \frac{n_i}{A} = -D_i \frac{dC_i}{dy} = -CD_i \frac{dx_i}{dy} \tag{1.10}$$

which is known as Fick's equation of diffusion. The substitution $C_i = Cx_i$, where x_i is the mole fraction of component i, is valid provided the molar density of the mixture C is constant. D_i is the diffusivity or diffusion coefficient of i in the mixture and has the dimensions L^2t^{-1}. Some values for D_i are listed in Table A.6.

The negative sign in equation (1.10) takes into account that diffusion occurs spontaneously from a region of high concentration to one of lower concentration. Again, we note the mathematical similarity of (1.10) to two previous equations, (1.6) and (1.8).

Example 1.4.3

The diffusion coefficient for carbon dioxide (A) in nitrogen (B) at 1 atm pressure and 20°C is given as 587 cm²/hr. Express this in the English engineering system of measurement.

Figure 1.3 Diffusion through a layer of fluid

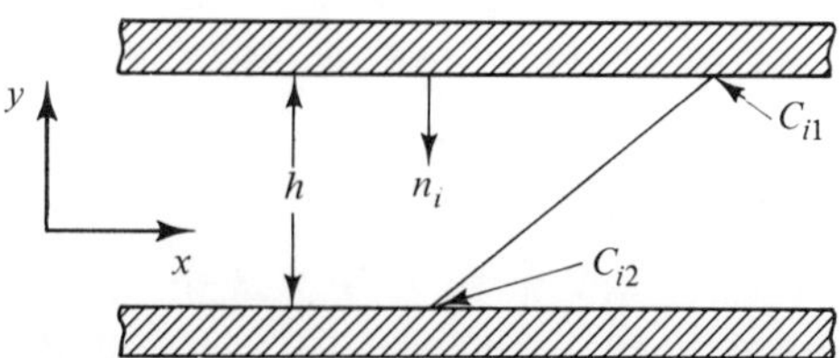

English engineering system:

$$D_{AB} = 587 \frac{\text{cm}^2}{\text{hr}} \times \left(\frac{\text{ft}}{12 \times 2.54 \text{ cm}}\right)^2 = 0.632 \text{ ft}^2/\text{hr}$$

1.5 Ratio of Diffusivities

In Section 1.4, we defined three different diffusivities, ν for momentum transfer, α for conductive heat transfer, and D for diffusive mass transfer. It is interesting that these three quantities, which relate to three different processes, all have the same dimensions, that is, L^2t^{-1}. It follows that the ratio of any two of these diffusivities must then be a number that has no dimensions. For instance, the ratio of ν and α gives a number that is known as the Prandtl number, Pr.

$$\text{Pr} \equiv \frac{\nu}{\alpha} = \frac{c\mu}{k}$$

Similarly, the ratio of α and D is the Lewis number, Le:

$$\text{Le} \equiv \frac{\alpha}{D} = \frac{k}{\rho c D}$$

And the ratio of ν and D is the Schmidt number, Sc:

$$\text{Sc} \equiv \frac{\nu}{D} = \frac{\mu}{\rho D}$$

The fact that these groups are dimensionless simply means that their numerical value is independent of any system of measurement. For instance, the Prandtl number for air at a particular temperature and pressure is a pure number that is the same in the SI, cgs, or any other system of measurement. We simply note these facts for now but will learn later that the special significance of these dimensionless groups lies in developing generalized correlations of data useful for predicting rates of momentum, heat, and mass transfer.

REFERENCES

1. Perry, R. H., and C. H. Chilton, *Chemical Engineer's Handbook*, 5th ed. New York: McGraw-Hill Book Company, 1973.
2. *Handbook of Chemistry and Physics.* Cleveland: Chemical Rubber Publishing Co., 1946.
3. Parrish, A., *Mechanical Engineer's Reference Book.* London: Butterworth & Co. Ltd., 1973.

PROBLEMS

1.1. Hydrogen is stored in metal containers at an absolute pressure of 1.8×10^4 kPa. What is the (a) mass density and (b) molar density of the hydrogen if the temperature of the container is 20°C? Express your answer in SI, metric, and English engineering units. *Answer:* (a) 15 kg/m^3, 15 g/l, 0.93 lb_m/ft^3; (b) 7.4 kg mol/m^3, 7.4 g mol/l, 0.46 lb mol/ft^3

1.2. The composition of dry air is approximately 79.0% N_2 and 21.0% O_2 by volume. Calculate (a) the molecular weight of air and (b) the density of air in grams/liter and in lb_m/ft^3 at 70°F and 741 mm Hg pressure. *Answer:* (a) 28.9; (b) 1.17 g/l, 0.0728 lb_m/ft^3

1.3. A more accurate composition of dry air than that given in Problem 1.2 is the following:

	% by volume
N_2	78.03
O_2	20.99
Ar	0.94
CO_2	0.031
Ne	0.009

(a) What is the molecular weight of air according to these data?
(b) Calculate the density of air in grams/liter and in lb_m/ft^3 at 70°F and 741 mm Hg.

Compare your answers to those obtained in Problem 1.2. *Answer:* (a) 28.96; (b) 1.17 g/l, 0.0730 lb_m/ft^3

1.4. A flue gas has the following composition by volume: 13.1% CO_2, 7.7% O_2, and 79.2% N_2. What is the molecular weight of the gas? *Answer:* 30.4

1.5. A vessel with a volume of 10 m^3 has nitrogen entering at a rate 5.0 kg/s and being discharged at the rate of 1.0 kg/s. If the vessel temperature is constant at 25°C, determine the rate at which the pressure rises inside the tank. *Answer:* 35 kPa/s

1.6. Oxygen is to be marketed in small cylinders each having a volume of 0.50 m^3 and containing 1.0 kg of the gas. If the cylinders may be subjected to a maximum temperature of 50°C, what must be their design pressure? *Answer:* 1.7×10^2 kPa

1.7. An automobile tire is inflated to a gauge pressure of 35 psi at a temperature of 0°F. What is the maximum temperature to which the tire may be heated without the gauge pressure exceeding 50 psi? Assume the volume of the tire remains constant. *Answer:* 139°F

1.8. A velocity gradient of 100 m/s-m is established in a fluid under the influence of a shear stress of 20.0 N/m^2. What is the dynamic viscosity of the fluid? *Answer:* 0.200 kg/m-s

1.9. A shaft with a diameter of 5.0 cm rotates inside a bearing at 700 rpm. Radial clearance between the shaft and the bearing surface is 0.045 mm, and the length of the bearing is 15 cm. The bearing is lubricated with an oil that has a viscosity of 0.010 kg/m-s. Determine the shear stress in the film of oil and the torque on the shaft. *Answer:* 407 N/m^2; 0.24 N-m

1.10. A circular disk 10 cm in diameter has a clearance of 0.30 mm from a flat plate. What torque is required to rotate the disk at 1800 rpm if the clearance space contains oil with a viscosity of 0.60 poise? *Answer:* 0.37 N-m

1.11. Water at 20°C flows over a flat plate. If the velocity profile at some position x on the plate is $v_x = 3y - y^3$ cm/sec, where y is the distance normal to the plate in cm, find the shear stress on the plate at the position x. *Answer:* 3.0×10^{-3} N/m^2

1.12. A shaft 2.500 cm in diameter is held in a journal bearing that is 30 cm long and 2.525 cm in diameter. Light oil with a viscosity of 0.023 kg/m-s at the temperature of the bearing is used as a lubricant. Calculate the power, expressed in SI units and as horsepower, that is required to rotate the shaft at 8000 rpm. *Answer:* 475 W or 0.64 hp

1.13. For the situation in Problem 1.11, suppose the temperature profile in the layer of water at position x is $T = 20 + 3 \sin(\pi/2\, y)$, where $0 < y < 1$, T is °C, and y is distance in cm. What is the heat flux through the plate at x? *Answer:* -2.8×10^2 W/m^2

1.14. The window in the living room of a house is 2.1 m × 2.8 m constructed of glass that is 6.0 mm thick. What is the rate of heat conduction through the window if the inside surface temperature is 20°C and the outside surface is (a) 40°C, (b) −10°C? Choose the positive x direction to be from inside to outside the house. *Answer:* (a) -1.5×10^4 W; (b) 2.3×10^4 W

1.15. A large plane slab of material 2.5 in. thick has a thermal conductivity of 0.005 cal/sec-cm-°C. To maintain a heat flux through the slab of 250 Btu/hr-ft^2, what temperature difference, in F°, must exist across the slab? *Answer:* 43F°

1.16. Determine the average thermal conductivity of a material that is 6.0 mm thick when it conducts 5 kW of energy per 1000 cm^2 of area when the temperature difference across the material is 40°C. *Answer:* 7.5 W/m °K

1.17. A layer of glass wool that is 5 in. thick has a temperature of 50°F on one side and 200°F on the other. Calculate the heat flux through the material in Btu/hr-ft^2. *Answer:* 15

1.18. What is the thermal diffusivity of liquid water at 20°C? *Answer:* 1.44×10^{-7} m^2/s

1.19. Air at 0°C and 1 atm pressure flows over a porous plate that is soaked in acetone. The molar concentration of acetone in the air at distance y from the plate is given by the following expression:

$$C_a = 5e^{-2y}, \qquad 0 < y < 1$$

where C_a = kg mol/m^3 and y = m. Calculate the flux of acetone from the plate. *Answer:* 0.393 kg mol/hr-m^2

1.20. The flux of water vapor from the surface of a large pool is 0.20 kg/hr-m^2 into dry air. If it is assumed that this occurs only by the process of diffusion, estimate the decrease in concentration of water vapor that must occur over a vertical distance of 10 cm from the air/water interface. *Answer:* 0.013 kg mol/m^3

Introduction to Fluid Mechanics

2

Fluid mechanics deals with the properties of fluids and the behavior of fluids at rest and in motion. Engineering applications of fluid mechanics include city water distribution to homes and factories, transport of oil and gas from distant oil fields, measurement of quantities of fluid flowing, lift and drag forces on structures, road vehicles, and aircraft, and many others. An introductory-level course in fluid mechanics provides sufficient background and insight to deal with many of these practical applications.

2.1 Fluid Properties

Density	$\rho(\text{kg/m}^3)$ or $(\text{lb}_\text{m}/\text{ft}^3)$, the mass of fluid occupying a unit volume
Specific volume	$v = \dfrac{1}{\rho}\,(\text{m}^3/\text{kg})$ or $(\text{ft}^3/\text{lb}_\text{m})$, the volume occupied per unit mass of fluid
Specific weight	$\gamma = \dfrac{g}{g_c}\,\rho(\text{N/m}^3)$ or $(\text{lb}_\text{f}/\text{ft}^3)$, the weight of fluid occupying unit volume
Specific gravity	$\text{SG} = \dfrac{\rho\text{ fluid}}{\rho\text{ water}} = \dfrac{\gamma\text{ fluid}}{\gamma\text{ water}}$, the ratio of fluid density to that of pure liquid water at 4°C

Dynamic viscosity — μ (N-s/m^2) or (lb$_f$-sec/ft^2), the proportionality constant in Newton's law of viscosity, $F/A = \mu\,\partial V/\partial y$, relating the viscous shear stress F/A to the rate of shear $\partial V/\partial y$

Kinematic viscosity — $\nu = \dfrac{\mu}{\rho}$ (m^2/s) or (ft^2/sec), the dynamic viscosity divided by the fluid density

2.2 Behavior of Fluids at Rest

A fluid may be a liquid or a gas. The defining characteristic of a fluid is its inability to resist shear forces when at rest. It follows that any surface submerged in a fluid at rest can experience normal forces only.

2.2.1 Fluid Pressure (N/m^2 = Pa) or (lb$_f$/ft^2). The pressure at a point in a fluid is one of the properties that is used to describe the state or condition of the fluid at that point. Consider a small plane surface, real or imaginary, placed in the fluid at the point where the pressure is to be evaluated. The pressure at the point is defined as $P = dF/dA$, where dF is the (normal) contact force between the fluid and either side of the small surface dA. The magnitude of dF and P is independent of the orientation of the small surface dA.

2.2.2 Variation of Pressure with Elevation in a Fluid. It is common knowledge that atmospheric air pressure decreases with height and that water pressure increases with depth. To deal with this familiar aspect of fluid behavior in a formal way, consider a small fluid element (Figure 2.1) that is part of a large stationary body of fluid. For the small fluid element to remain stationary, the forces acting on it must balance.

$$\sum F_Z = 0: \quad P\,dA - (P + dP)dA - \frac{\rho g}{g_c}\,dz\,dA = 0$$

Hence

$$dP = -\frac{\rho g\,dz}{g_c} = -\gamma\,dz \tag{2.1}$$

This is a fundamental equation for fluids at rest in a gravitational field. Note that, in the general case, γ may be a variable over the domain under consideration. In the particular case of pressure in a liquid, γ can be treated as a constant and, if desired, the negative sign in equation (2.1) can be eliminated by treating the downward direction as positive. Representing depth by the symbol h, we may rewrite (2.1) for liquids:

$$dP = +\gamma\,dh$$

$$P = \gamma h + C$$

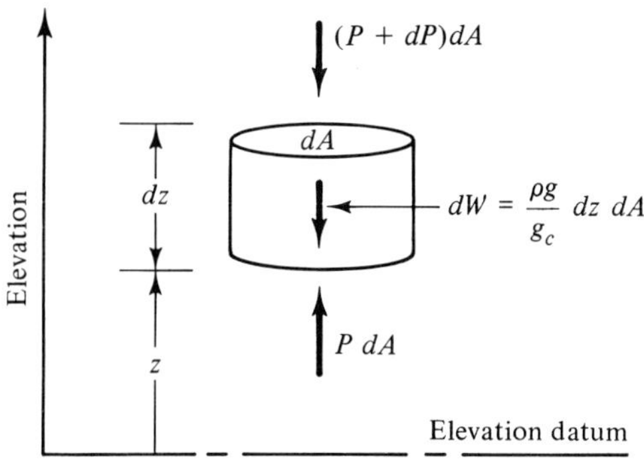

Figure 2.1 Forces acting on fluid element

At $h = 0$ (liquid surface), $C = P$ at surface.

Thus

$$P_{\text{at depth } h} - P_{\text{at surface}} = P_{\text{gauge}} = \gamma h \tag{2.1a}$$

2.2.3 Example Problems

1. Determine the gauge and absolute pressure in water at 16°C, 100 m below the water surface. Atmospheric air pressure at the surface is 101 kPa.

   Solution:

$$P_{\text{gauge at 100 m}} = \gamma h = (999 \times 9.807 \times 100) = 979{,}700 \text{ N/m}^2$$

$$= 980 \text{ kPa}$$

$$P_{\text{absolute at 100 m}} = P_{\text{atmos}} + P_{\text{gauge}} = 101 + 980 = 1081 \text{ kPa}$$

2. Determine the absolute air pressure at an altitude of 2000 m above sea level assuming that the air temperature is constant at 20°C.

   Solution:

$$\text{air density } \rho = \frac{PM}{RT}$$

Thus

$$dP = -\gamma\, dz = -\rho g\, dz = -\frac{PM}{RT} g\, dz \qquad (\textit{Note: } g_c = 1 \text{ in SI})$$

and

$$\int_{P_0}^{P} \frac{dP}{P} = -\frac{Mg}{RT}\int_{z_0}^{z} dz = -\frac{28.9 \times 9.807 \text{ (m/s}^2)}{8314 \text{ (N m/kg mol K)} \times 293 \text{ K}} \int_{z=0}^{z=2000} dz$$

$$\ln \frac{P}{P_0} = -0.2327 \quad \text{(dimensionless)}$$

Thus

$$\frac{P}{P_0} = e^{-0.2327} = 0.7924$$

$$\text{Absolute air pressure at 2000 m} = 0.792 \times P_0 = 80.0 \text{ kPa}$$

2.2.4 Manometers for Pressure Measurement. A manometer is a transparent tube containing a liquid. Based on the known relation between pressure and depth in a liquid ($\Delta P = \gamma h$), the liquid height can be used to indicate pressure or pressure difference. Various tube configurations can be used depending on the specific application. Several manometer configurations are illustrated in Figure 2.2

Consider a fluid having specific weight γ flowing through a pipe.

1. We can determine the gauge pressure at point A by allowing fluid to rise in a vertical open tube:

$$P_A = P_{\text{atmos}} + \gamma h_1$$

$$P_{A\ \text{gauge}} = P_A - P_{\text{atmos}} = \gamma h_1$$

2. It may be that the required height h_1 in Figure 2.2(a) is too large for convenience. An alternative is to use a second manometer fluid such as mercury having specific weight γ_M arranged as shown in Figure 2.2(b).

$$P_B + \gamma h_2 - \gamma_M h_3 = P_{\text{atmos}}$$

$$P_{B\ \text{gauge}} = P_B - P_{\text{atmos}} = \gamma_M h_3 - \gamma h_2$$

3. We may wish to know the pressure difference $P_C - P_D$ as the fluid in the pipe flows past an obstacle. Using the manometer arrangement

Figure 2.2 Several manometer configurations

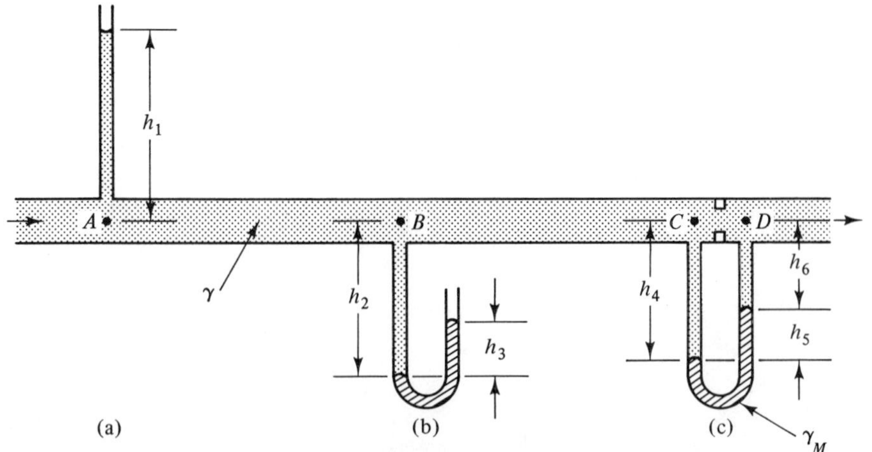

shown in Figure 2.2(c),

$$P_C + \gamma h_4 - \gamma_M h_5 - \gamma h_6 = P_D$$

Note that $h_4 - h_6 = h_5$.

$$P_C - P_D = \gamma_M h_5 - \gamma(h_4 - h_6) = (\gamma_M - \gamma)h_5$$

2.2.5 Mercury Barometer. Most problems in incompressible flow require knowledge of gauge pressures only, and these can be measured using manometers or pressure gauges. In some cases, such as when using $P = \rho RT/M$ for determining the properties of gases, we have to know the absolute pressure, which implies knowing the atmospheric pressure as well as the gauge pressure. Using a mercury barometer (Figure 2.3), the atmospheric pressure can be calculated.

$$P_{\text{vacuum}} + \gamma_M h = P_{\text{atmos}}$$

$$\therefore \quad P_{\text{atmos}} = \gamma_M h$$

Assuming a column height of 760 mm Hg (standard atmosphere) and given the specific gravity of mercury, $SG_M = 13.59$,

$$P_{\text{std atmos}} = (999 \times 9.807 \times 13.59) \times 0.760$$

$$= 101000 \text{ N/m}^2 = 101 \text{ kPa}$$

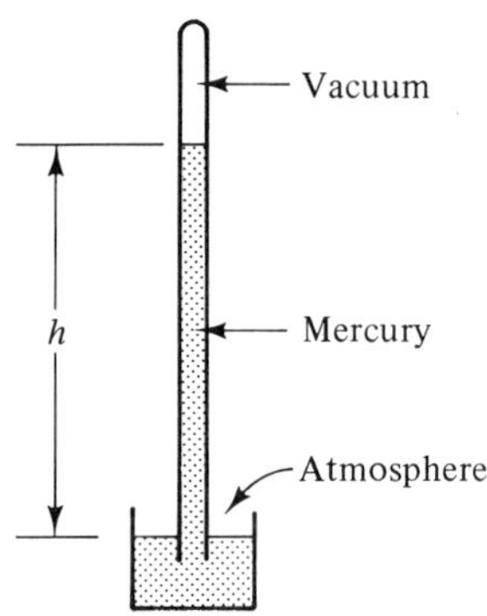

Figure 2.3 Barometer

2.3 Description of Flow

2.3.1 Steady and Unsteady Flow. Flow is *steady* when conditions do not change with time at any location in the space occupied by fluid. Thus at any location in the fluid $\partial V/\partial t = 0$, $\partial P/\partial t = 0$, $\partial \rho/\partial t = 0$, and so on.

The flow is *unsteady* when conditions at any location change with time. While a valve in a pipeline is being opened, and the fluid velocity in the pipe is increasing, the flow is unsteady. Once the valve is open and the fluid velocity has settled at a constant value, the flow is steady.

2.3.2 One-, Two-, and Three-Dimensional Flow. Three-dimensional flow is the most general case, involving velocity components and pressure gradients in the x, y, and z directions. A more restricted case, such as frictionless flow between parallel sheets of glass, would involve velocity components and pressure gradients in two directions, x and y, and is called two-dimensional flow. In one-dimensional flow the important changes occur in the direction of flow. Minor changes perpendicular to the flow direction are ignored.

2.3.3 Streamlines. A *streamline* is defined as a continuous line drawn through the fluid so that at every point it is parallel to the velocity vector at that point. Since there is no velocity normal to the streamline, no fluid mass can cross a streamline. A short time-exposure photograph of a two-dimensional flow field sprinkled with particles will produce short dashes indicating the flow direction at every point in the field (similar to the iron filings experiment with a magnet). Streamlines can now be sketched, everywhere parallel to the short dashes.

2.3.4 Laminar and Turbulent Flow. In *laminar flow*, fluid particles move in smooth paths with one layer gliding smoothly over an adjacent layer.

In *turbulent flow*, fluid particles move with an irregular tumbling motion. Strictly speaking, turbulent flow in pipes is unsteady since local velocities fluctuate with time. However, the average stream velocity can still be treated as steady.

Reynolds demonstrated the distinction between laminar and turbulent flow characteristics by injecting dye into a flowing fluid (Figure 2.4). He also showed that the nature of the flow can be correlated with a dimensionless group called the Reynolds number $\text{Re} = DV\rho/\mu$. For fluid flow in pipes, the flow will be laminar for $\text{Re} < 2000$, laminar or turbulent in the range $2000 < \text{Re} < 4000$, and generally turbulent for $\text{Re} > 4000$. Because the Reynolds number is dimensionless, the numerical value of Re is the same, regardless of the system of units employed.

$$\text{Re} = \frac{\rho(\text{lb}_\text{m}/\text{ft}^3) \times V\,(\text{ft/sec}) \times D\,(\text{ft})}{\mu(\text{lb}_\text{m}/\text{sec ft})} = \frac{\rho(\text{kg/m}^3) \times V\,(\text{m/s}) \times D\,(\text{m})}{\mu(\text{kg/s-m})}$$

2.3.5 Mathematical Description of Fluid Flow. The Lagrangian method is to trace the motion of individual fluid particles and the forces acting on the particles. While this is consistent with the method used in particle mechanics, it is rarely suitable for practical application because of the difficulty of tracing individual fluid particle motion in a flow field.

The Eulerian method describes the velocity and pressure at fixed points in the flow field as, for instance, at the entrance to a pipe or at a pipe elbow.

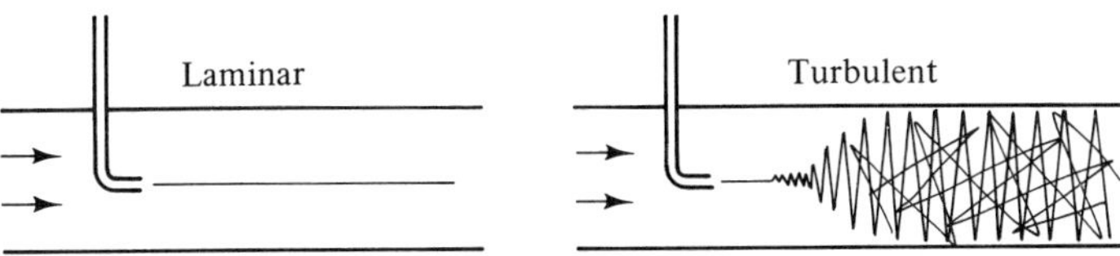

Figure 2.4 Reynold's experiment

It is the method commonly used since standard instruments for measuring pressure or velocity are installed in fixed locations.

In general, fluid particle velocity depends on location and time. For example, consider the fluid velocity in a converging duct as illustrated in Figure 2.5. If the flow rate is steady, the fluid velocity measured by a sensor at A will remain constant. However, the fluid particle passing A is moving from a low-velocity region to a high-velocity region. That is, the fluid particle passing point A has a *convective* acceleration due to its changing location in the field.

$$a_{\text{convective}} = V\frac{\partial V}{\partial s}$$

Now imagine that the volume flow rate through the pipe is increasing with time. A velocity sensor at A would measure a *local* increase of velocity at A due to the increasing flow rate. A fluid particle passing point A would have a *local* acceleration because the fluid field velocity at location A is increasing with time.

$$a_{\text{local}} = \frac{\partial V}{\partial t}$$

The total (*substantial* or *material*) acceleration of a particle, which is the sum of the convective and local accelerations, can be shown more succinctly by the use of partial differentiation. The fluid velocity is a function of location and time:

$$V = V(s, t)$$

Then

$$\frac{dV}{dt} = \frac{\partial V}{\partial s} \times \frac{ds}{dt} + \frac{\partial V}{\partial t} \times \frac{dt}{dt} \tag{2.2}$$

Figure 2.5 Flow in a converging duct

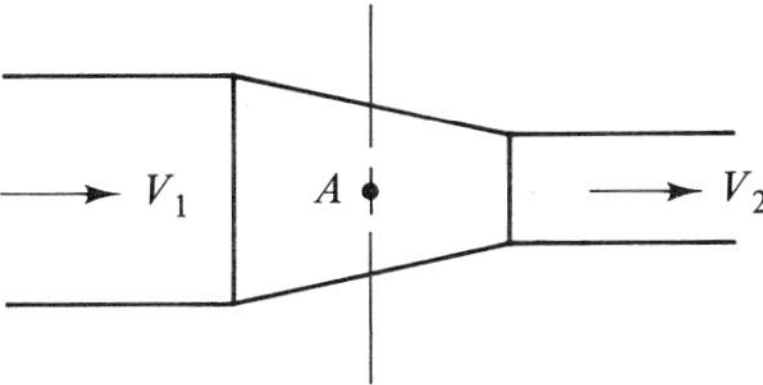

or

$$\frac{dV}{dt} = V\frac{\partial V}{\partial s} + \frac{\partial V}{\partial t}$$

2.4 Equation of Continuity (Conservation of Mass)

Consider flow along a stream filament, enclosed within a family of streamlines between sections 1 and 2, as illustrated in Figure 2.6. The principle of mass conservation dictates the following:

$$\begin{pmatrix}\text{rate of mass}\\ \text{entering}\\ \text{through } A_1\end{pmatrix} = \begin{pmatrix}\text{rate of mass}\\ \text{accumulation in}\\ \text{enclosed volume}\end{pmatrix} + \begin{pmatrix}\text{rate of mass}\\ \text{leaving}\\ \text{through } A_2\end{pmatrix}$$

or

$$\rho_1 A_1 V_1 = \text{vol} \times \left(\frac{d\rho}{dt}\right)_{\text{average}} + \rho_2 A_2 V_2 \tag{2.3}$$

In the case of steady flow, the local density does not change with time, and there is no mass accumulation within the volume. Thus, the equation of continuity for steady flow takes the form

$$\rho_1 A_1 V_1 = \rho_2 A_2 V_2 \tag{2.4}$$

The mass flow rate across any section of the stream filament is the same as across any other section.

$$\dot{m} = \rho A V = \text{constant} \tag{2.5}$$

For flow through a pipe (or through an imaginary bundle of stream tubes), the velocity may vary across the stream section. The mass flow rate across a section is given by

$$\dot{m} = \int_A \rho V \, dA = \rho A \overline{V} \tag{2.6}$$

where $\overline{V}$ = average velocity at that cross section.

Figure 2.6 Flow through a stream filament

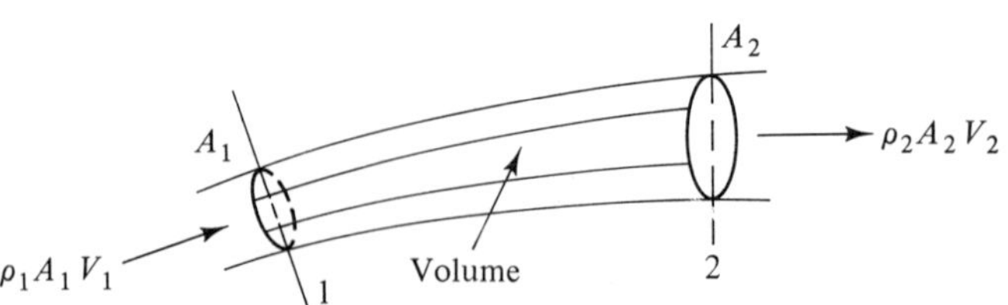

For incompressible flow, $\rho_1 = \rho_2$, and the volume rate of flow can be expressed as

$$Q = A_1V_1 = A_2V_2 = AV = \text{constant} \tag{2.7}$$

2.5 Relationship between Pressure and Velocity

Whether dealing with fluid flow through pipes or flow around submerged objects, the relationship between pressure and velocity is generally required. A description of this relationship over the whole field of flow may be complicated, but it is relatively simple along a single streamline. Many engineering problems may be solved, with some approximations, using this simple pressure–velocity relationship along a streamline.

Consider the motion of a prismatic fluid element along a streamline (Figure 2.7). Apply Newton's second law to the element:

$$\sum F_s = \frac{m}{g_c}\frac{dV}{dt} \tag{2.8}$$

$$P\,dA - (P + \frac{\partial P}{\partial s}ds)dA - dW\cos\phi = \frac{\gamma\,dA\,ds}{g}\frac{dV}{dt}$$

Note that $V = V(s, t)$.

$$\frac{dV}{dt} = V\frac{\partial V}{\partial s} + \frac{\partial V}{\partial t}$$

Also, $dW = \gamma dA\,ds$ and $\cos\phi = \partial z/\partial s$. Thus

$$-\frac{\partial P}{\partial s}ds\,dA - \gamma\,dA\,ds\frac{\partial z}{\partial s} = \frac{\gamma\,dA\,ds}{g}\left(V\frac{\partial V}{\partial s} + \frac{\partial V}{\partial t}\right)$$

Figure 2.7 Fluid motion along a streamline

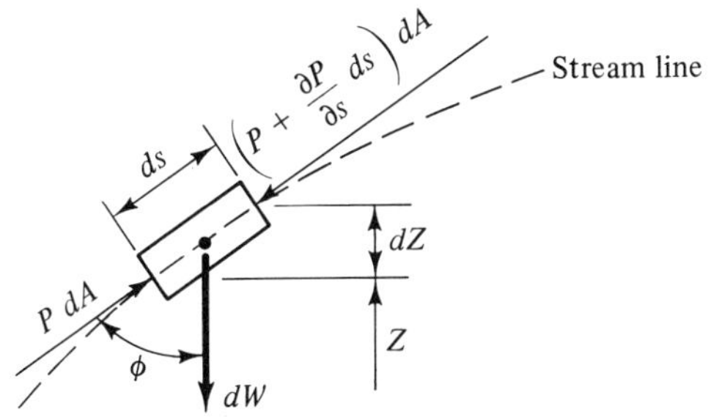

Divide by $\gamma\, dA\, ds$ and change sign:

$$\frac{1}{\gamma}\frac{\partial P}{\partial s} + \frac{\partial z}{\partial s} + \frac{V}{g}\frac{\partial V}{\partial s} = -\frac{1}{g}\frac{\partial V}{\partial t} \tag{2.9}$$

This is Euler's equation for frictionless flow along a streamline. In steady flow, $\partial V/\partial t = 0$; hence

$$\frac{1}{\gamma}\frac{\partial P}{\partial s} + \frac{V}{g}\frac{\partial V}{\partial s} + \frac{\partial z}{\partial s} = 0$$

Because s is now the only independent variable, we can substitute the total derivative for the partial derivative:

$$\frac{1}{\gamma}\frac{dP}{ds} + \frac{V}{g}\frac{dV}{ds} + \frac{dz}{ds} = 0 \tag{2.10}$$

Thus *Euler's equation for steady flow* along a streamline can be expressed as follows, after multiplying through by ds.

$$\frac{dP}{\gamma} + \frac{V\,dV}{g} + dz = 0 \tag{2.11}$$

Integration of Euler's equation for *incompressible flow* yields *Bernoulli's equation* for frictionless flow along a streamline.

$$\frac{P}{\gamma} + \frac{V^2}{2g} + z = \text{constant} \tag{2.12}$$

Thus the pressure, velocity, and elevation at two separated points along a streamline are related by

$$\frac{P_1}{\gamma} + \frac{V_1^{\,2}}{2g} + z_1 = \frac{P_2}{\gamma} + \frac{V_2^{\,2}}{2g} + z_2 \tag{2.13}$$

As expressed here, each of the terms in Bernoulli's equation represents energy per unit weight of fluid [dimensions of (force × length)/force].

$\dfrac{P}{\gamma}$ flow work as defined in thermodynamics

work done by the pressure as unit weight of fluid crosses a particular location

$\dfrac{V^2}{2g}$ kinetic energy content of unit weight of fluid

z potential energy of unit weight of fluid

The reader may have noticed that the force dimension could be canceled out of the numerator and denominator of units for each of the Bernoulli terms, leaving just the length dimension. This gives rise to another convenient interpretation of the Bernoulli terms (Figure 2.8) in which each term in the Bernoulli equation can be represented by a height of fluid.

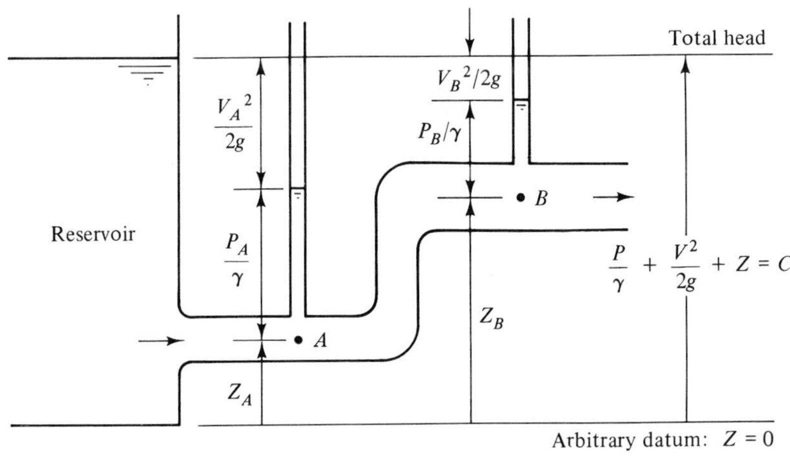

Figure 2.8 Bernoulli terms in equation (2.13)

z elevation of fluid element in gravitational field measured from an arbitrary datum

$\frac{P}{\gamma}$ fluid column height that would be supported by the fluid pressure in the flowing stream

$\frac{V^2}{2g}$ fluid column height associated with the fluid velocity; the vertical distance ($V_A^2/2g$) between the free liquid surface in the column above A and the liquid surface level in the reservoir results from pressure having been "used" to create the kinetic energy at A

Another presentation of Bernoulli's equation is in terms of energy per unit mass. Substitute for γ the expression $\rho g/g_c$ and then multiply all the terms of Bernoulli's equation by (g/g_c) to get

$$\frac{P}{\rho} + \frac{V^2}{2g_c} + g\frac{z}{g_c} = \text{constant} \tag{2.14}$$

In the SI system of units, g_c is unity and dimensionless, so equation (2.14) becomes

$$\frac{P}{\rho} + \frac{V^2}{2} + gz = \text{constant} \quad \text{(units are N-m/kg)} \tag{2.15}$$

Strictly speaking, Bernoulli's equation is valid only along a single streamline. However, for frictionless flow originating in a reservoir, the Bernoulli constant (total head) for each streamline is the same, and the Bernoulli relation can therefore be used to describe the bulk flow in a pipe.

2.6 Linear Momentum Equation

The impulse–momentum principle provides another independent relation between fluid forces, pressures, and velocities. Consider the flow of a fluid through a defined volume in space (Figure 2.9).

At time t

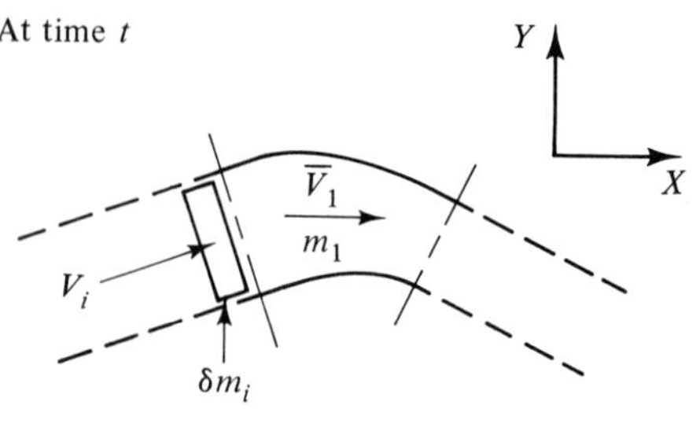

At time $(t + \delta t)$

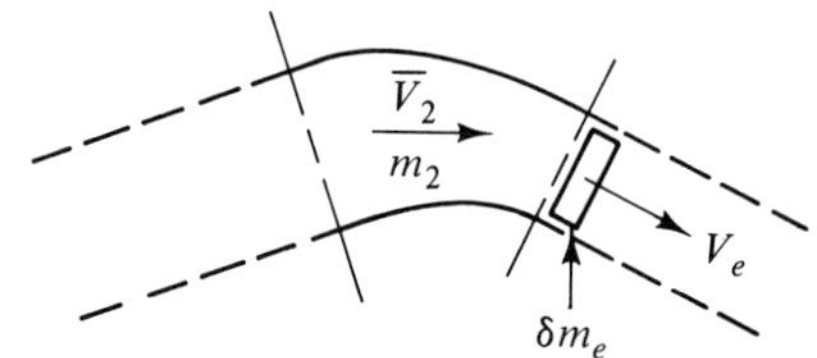

Figure 2.9 Flow through a defined volume

m fluid mass within the defined volume in space
$\overline{V}$ average velocity of m
V_i fluid velocity entering
V_e fluid velocity leaving
$\dot{m}_i$ mass flow rate entering
$\dot{m}_e$ mass flow rate leaving

We will analyze the effect of all external forces acting on a fixed quantity of mass. This fixed quantity of mass includes $m_1 + \delta m_i$ at time t, and this same mass is $m_2 + \delta m_e$ at time $(t + \delta t)$.

Applied impulse during δt

$$= \text{final momentum} - \text{initial momentum:} \tag{2.16}$$

$$\sum F \times \delta t = \frac{1}{g_c}[(m_2\overline{V}_2 + \delta m_e V_e) - (m_1\overline{V}_1 + \delta m_i V_i)]$$

$$= \frac{1}{g_c}[(m_2\overline{V}_2 - m_1\overline{V}_1) + \delta m_e V_e - \delta m_i V_i]$$

$$= \frac{1}{g_c}[\delta(m\overline{V}) + \delta m_e V_e - \delta m_i V_i]$$

Divide by δt and replace the difference δ by the differential

$$\sum F = \frac{1}{g_c}\left[\frac{d}{dt}(m\overline{V}) + \dot{m}_e V_e - \dot{m}_i V_i\right] \tag{2.17}$$

For steady flow

$$\frac{d}{dt}(m\overline{V}) = 0 \quad \text{and} \quad \dot{m}_e = \dot{m}_i = \dot{m}$$

Thus the linear momentum equation for steady flow is

$$\sum F = \frac{\dot{m}}{g_c}(V_e - V_i) \tag{2.18}$$

This is a vector equation and is usually applied to one component at a time.

$$\sum F_x = \frac{\dot{m}}{g_c}(V_{e_x} - V_{i_x}) \tag{2.19a}$$

$$\sum F_y = \frac{\dot{m}}{g_c}(V_{e_y} - V_{i_y}) \tag{2.19b}$$

If we just memorize the momentum equation, we will be limited in our ability to handle a variety of problem types. The momentum equation for steady flow is more powerful when expressed in words:

$$\begin{pmatrix}\text{Net applied}\\ \text{force}\end{pmatrix} = \begin{pmatrix}\text{Rate of momentum}\\ \text{leaving}\end{pmatrix} - \begin{pmatrix}\text{rate of momentum}\\ \text{entering}\end{pmatrix} \tag{2.20}$$

This word equation can be applied to systems involving several streams entering and leaving.

2.7 Flow Measurement

We have now derived the three basic equations for steady one-dimensional flow.

Continuity $\quad \dot{m} = \rho_1 A_1 V_1 = \rho_2 A_2 V_2 = \rho A V = \text{constant} \quad$ (2.4)

Bernoulli $\quad \dfrac{P_1}{\gamma} + \dfrac{V_1^2}{2g} + z_1 = \dfrac{P_2}{\gamma} + \dfrac{V_2^2}{2g} + z_2 \quad$ (2.13)

Momentum $\quad \sum F = \dfrac{\dot{m}}{g_c}(V_e - V_i) \quad$ (2.18)

In the next chapter on applications in fluid flow, we will introduce some of the practical consequences of viscosity and friction and, in particular, the modifications required to the Bernoulli equation. However, there are several real flow-measuring devices whose performance is nearly ideal. Analysis of these devices, important for their engineering uses, will also serve as practice in the application of the basic concepts and equations.

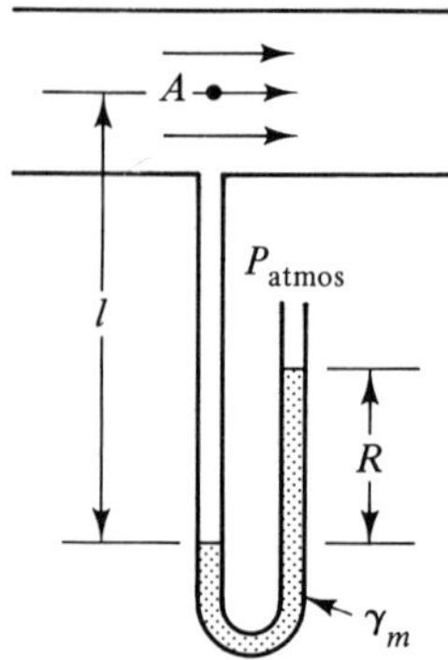

Figure 2.10 U-tube manometer [similar to Figure 2.2(b)]

2.7.1 Measurement of Static, Stagnation, and Dynamic Pressures. The *static pressure* at a point in a fluid is simply the pressure at that point. In Figure 2.10, a manometer is used to sense this pressure. If a force balance on the manometer is written starting at point A, the following is obtained.

$$P_A + \gamma l = P_{atmos} + \gamma_m R$$

or (2.21)

$$P_{A\ absolute} = P_{atmos} + \gamma_m R - \gamma l$$

where $P_{A\ absolute}$ is the absolute pressure at A.

The gauge pressure at A is defined as follows:

$$P_{A\ gauge} = P_{A\ absolute} - P_{atmos} = \gamma_m R - \gamma l \tag{2.22}$$

The *stagnation pressure* in a moving fluid, at point B in Figure 2.11(a), is the pressure that would exist if the fluid were brought to rest in a frictionless process. It can be sensed with an impact tube as illustrated in Figure 2.11(b). The stream filament traveling from A to B is assumed to come to rest at the opening of the impact tube. Applying the Bernoulli equation between points A and B, as well as between A and B_s gives the following:

$$\frac{P_A}{\gamma} + \frac{V_A^2}{2g} = \frac{P_B}{\gamma} + \frac{V_B^2}{2g} \tag{2.23}$$

$$\frac{P_A}{\gamma} + \frac{V_A^2}{2g} = \frac{P_{BS}}{\gamma} + \frac{V_{BS}^2}{2g} \tag{2.24}$$

Comparing (2.23) and (2.24), and noting that $V_{BS} = 0$, gives

$$P_{BS} = P_B + \frac{\gamma V_B^2}{2g} \tag{2.25}$$

The term P_{BS} is the stagnation pressure at B, and the term $\gamma V_B^2/2g$ (or $\rho V_B^2/2g_c$) is called the *dynamic pressure*.

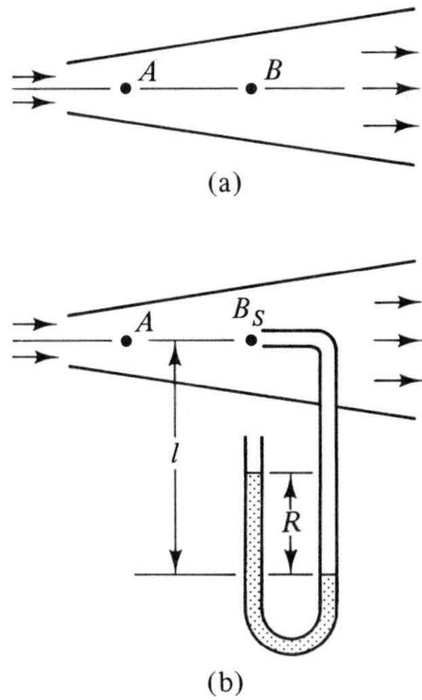

Figure 2.11 Impact tube

2.7.2 Measurement of Velocity Using a Pitot-Static Tube.

The Pitot-static tube illustrated in Figure 2.12 has two openings, at *B* and *C*, connected across a simple U-tube manometer. The opening at *B* senses the stagnation pressure, and the opening at *C* senses the static pressure. From equation (2.25),

$$P_{BS} - P_C = \frac{\gamma V_B^2}{2g}$$

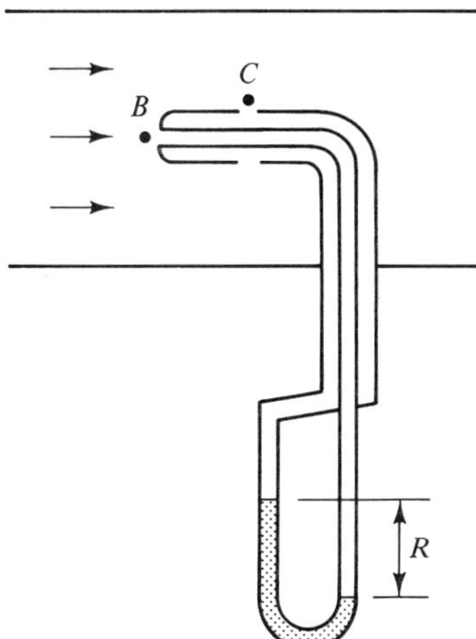

Figure 2.12 Pitot tube

Now using the manometer

$$P_{BS} - P_C = R(\gamma_m - \gamma)$$

Thus, the fluid velocity in the vicinity of point B is given by

$$V_B = \left[2gR\left(\frac{\gamma_m}{\gamma} - 1\right)\right]^{1/2} \tag{2.26}$$

2.7.3 Venturi Meter and Flow Nozzle. If it is not convenient to measure flow rate by weighing and timing, a secondary method may be used, such as a venturi meter (Figure 2.13), nozzle (Figure 2.14), or orifice (Figure 2.16).

A venturi is a converging–diverging passage built into a pipe. The static pressure decreases from inlet to throat as the velocity increases, and this pressure change can be correlated with the flow rate. The venturi diffuser, which is the diverging section, reduces the fluid velocity and converts some of the dynamic pressure at the throat back to static pressure at the exit.

Flow rate through a nozzle is measured and calculated in the same manner as flow through a venturi. However, without the diverging section, the recovery of dynamic pressure from throat to downstream pipe is not as efficient as in the venturi.

Consider the pressure–velocity relationship between sections 1 and 2 (in either venturi or nozzle) as given by Bernoulli's equation.

$$\frac{P_1}{\gamma} + \frac{V_1^2}{2g} = \frac{P_2}{\gamma} + \frac{V_2^2}{2g}$$

From continuity, $Q = A_1V_1 = A_2V_2$ for an incompressible fluid. Thus

$$\frac{V_2^2 - V_1^2}{2g} = \frac{V_2^2}{2g}\left[1 - \left(\frac{A_2}{A_1}\right)^2\right] = \frac{P_1 - P_2}{\gamma}$$

Figure 2.13 Venturi meter

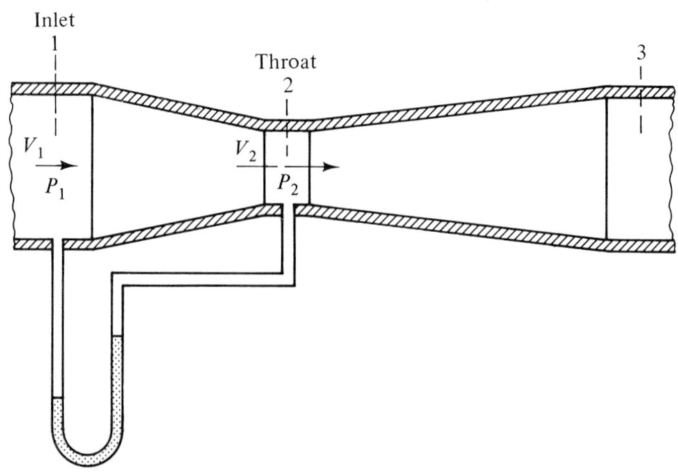

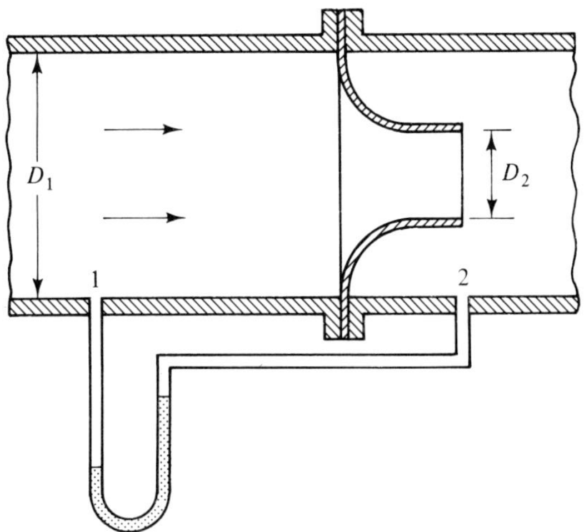

Figure 2.14 Nozzle

and

$$V_2 = \sqrt{\frac{2g(P_1 - P_2)/\gamma}{1 - \left(\frac{A_2}{A_1}\right)^2}} \tag{2.27}$$

The velocity of approach factor $\sqrt{1 - (A_2/A_1)^2}$ is normally expressed as $\sqrt{1 - \beta^4}$, where β = diameter ratio D_2/D_1. Thus the ideal volume flow rate is given by

$$Q_{\text{ideal}} = A_2V_2 = \frac{A_2}{\sqrt{1 - \beta^4}} \sqrt{2g\,\frac{P_1 - P_2}{\gamma}} \tag{2.28}$$

The actual volume flow rate differs from the ideal because of friction and the fact that the upstream velocity profile is not uniform. For a given flow-measuring device, the difference between the ideal and actual flow rates is accounted for by use of an empirical discharge coefficient C, which is a function of the Reynolds number.

$$Q_{\text{actual}} = CA_2V_2 = \frac{CA_2}{\sqrt{1 - \beta^4}} \sqrt{2g\,\frac{P_1 - P_2}{\gamma}} \tag{2.29}$$

For best results, the discharge coefficient should be determined experimentally in the actual installation. Failing this, agreed upon coefficients can be determined from the publication [1].

This discharge coefficient for a standard venturi differs slightly from that for a standard nozzle. However, we will use the typical nozzle discharge coefficient curve of Figure 2.15 for all our nozzle and venturi problems.

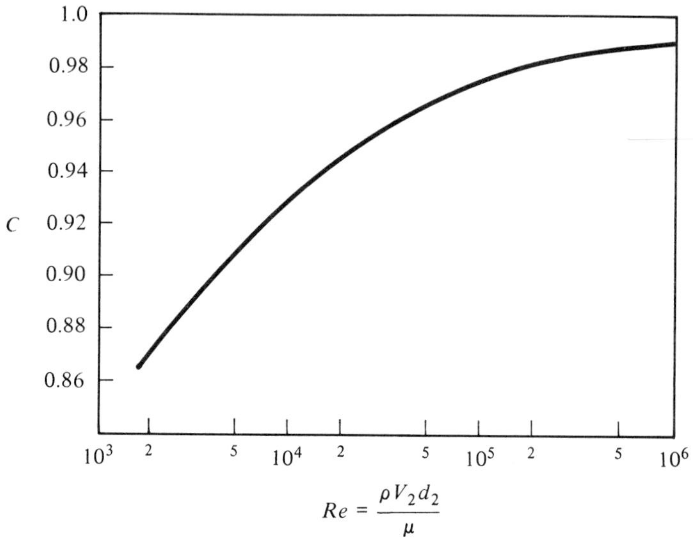

Figure 2.15 Nozzle discharge coefficient

2.7.4 Orifice Meter. The orifice meter illustrated in Figure 2.16 is essentially a plate with a clean-cut sharp-edged hole installed between two flanges in the pipeline. The restriction caused by the orifice creates a pressure drop that is used to measure flow rate. The upstream fluid must curve toward the pipe centerline to pass through the orifice, and, because of inertia, the jet continues to contract beyond the orifice opening. The minimum area of the jet (point 2 in Figure 2.16) is called the *vena contracta*. The vena contracta area A_2 is the orifice area A_o multiplied by a factor called the *contraction coefficient*, C_c. Note that the flow to the vena contracta can be treated as

Figure 2.16 Orifice meter

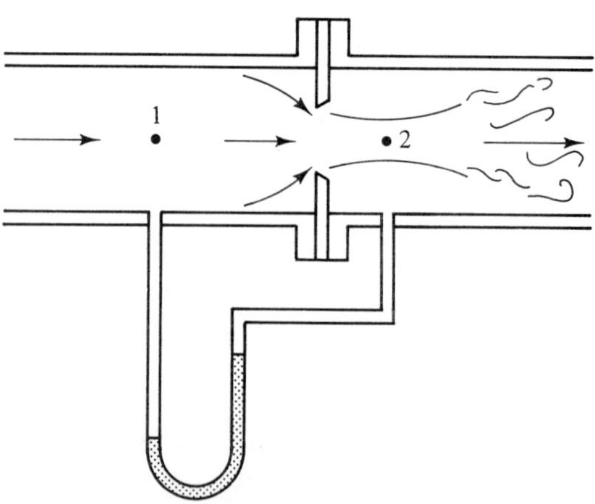

nearly frictionless, but that beyond the vena contracta the turbulent mixing of the jet is a frictional type of process that converts kinetic energy to thermal energy.

Applying Bernoulli's equation from point 1 to 2 gives

$$\frac{P_1}{\gamma} + \frac{V_1^2}{2g} = \frac{P_2}{\gamma} + \frac{V_2^2}{2g}$$

or

$$\frac{V_2^2 - V_1^2}{2g} = \frac{P_1 - P_2}{\gamma}$$

Continuity gives $Q = A_1V_1 = A_2V_2 = C_cA_oV_2$ for an incompressible fluid. Thus

$$\left(\frac{V_1}{V_2}\right)^2 = \left(\frac{A_2}{A_1}\right)^2 = \left(\frac{C_cA_o}{A_1}\right)^2 = C_c^2\beta^4$$

$$\frac{V_2^2}{2g}(1 - C_c^2\beta^4) = \frac{P_1 - P_2}{\gamma}$$

$$V_{2\ \text{frictionless}} = \frac{1}{\sqrt{1 - C_c^2\,\beta^4}}\sqrt{2g\,\frac{P_1 - P_2}{\gamma}}$$

The actual average velocity at the vena contracta is slightly reduced because of viscous forces acting between the upstream face of the orifice plate and the fluid. This velocity reduction can be accounted for by a velocity coefficient C_v, giving

$$V_2 = \frac{C_v}{\sqrt{1 - C_c^2\,\beta^4}}\sqrt{2g\,\frac{P_1 - P_2}{\gamma}} \tag{2.30}$$

The flow rate is given by

$$Q_{\text{orifice}} = A_2V_2 = C_cA_oV_2 \tag{2.31}$$

$$= \frac{C_vC_cA_o}{\sqrt{1 - C_c^2\,\beta^4}}\sqrt{2g\,\frac{P_1 - P_2}{\gamma}}$$

In practice, it is difficult to distinguish between the effects of C_c and C_v, so both effects are lumped together into a single orifice discharge coefficient C.

$$Q_{\text{orifice}} = \frac{CA_o}{\sqrt{1 - \beta^4}}\sqrt{2g\,\frac{P_1 - P_2}{\gamma}} \tag{2.32}$$

For Reynolds numbers ($\rho V_oD_o/\mu$) greater than 30,000, the orifice discharge coefficient is substantially constant at $C \simeq 0.61$.

An alternative procedure is to group the discharge coefficient C and the velocity of approach factor $\sqrt{1 - \beta^4}$ together in a single factor K called the *flow coefficient* (see figure 2.17). Either equation (2.32) or (2.33) may be used to solve orifice flow problems.

$$Q_{\text{orifice}} = KA_o \sqrt{2g \frac{P_1 - P_2}{\gamma}} \tag{2.33}$$

2.7.5 Overall Head Loss Due to Nozzle in Pipe. To calculate flow rate through nozzles, venturis and orifices, the continuity and Bernoulli's equations were required but not the linear momentum equation. The following example requires all three basic equations.

Figure 2.18 illustrates the static and total pressure variation adjacent to a nozzle (expressed in terms of head $= P/\gamma$). The purpose of the nozzle constriction is to cause a velocity increase so that the accompanying reduction in static pressure at point 2 can be used as a measure of mass flow rate. The overall head loss associated with the flow measurement is important in terms

Figure 2.17 Orifice coefficients (Streeter, V. L., *Fluid Mechanics*, © 1958, p. 321. Adapted by permission of McGraw-Hill Book Company, New York)

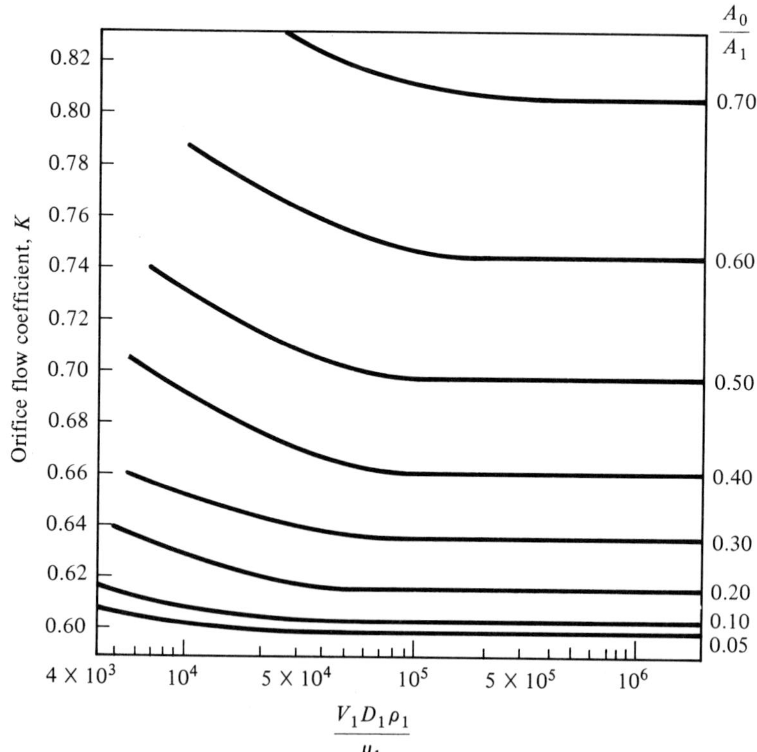

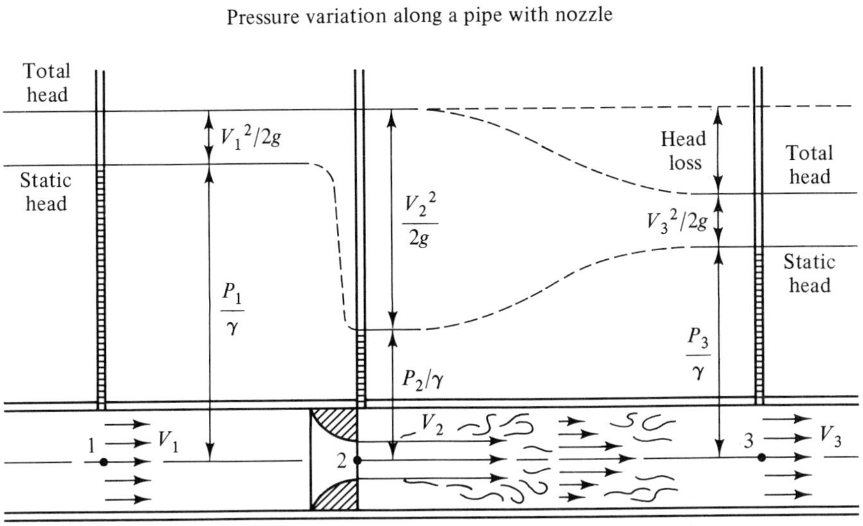

Pressure variation along a pipe with nozzle

Figure 2.18 Pressure variation along a pipe with nozzle

of the pump work required to transport the fluid through the pipe. The overall head loss is the decrease in total pressure from the uniform condition at point 1 to the uniform condition at point 3. In this case the head loss is also equal to the drop in static head from point 1 to 3 because the dynamic heads ($V^2/2g$) are equal at points 1 and 3.

We can use the Bernoulli equation for the pressure–velocity relation from point 1 to 2 as the flow is smooth and relatively friction free.

$$\frac{P_1 - P_2}{\gamma} = \frac{V_2^2 - V_1^2}{2g} \tag{2.34}$$

We cannot use the Bernoulli equation from point 2 to 3 as the flow is a mixing, frictional-type process. The linear momentum equation provides the required pressure–velocity relation from point 2 to 3. Figure 2.19 is a free-body diagram indicating the forces acting on the fluid between points 2 and 3. Equating the net force in the x direction to the change in momentum gives the following:

$$\sum F_x = \frac{\dot{m}}{g_c}(V_3 - V_2)_{x \text{ direction}}$$

$$(P_2 - P_3)A_3 = \frac{\rho}{g_c} A_3 V_3 (V_3 - V_2) \tag{2.35}$$

Since $\rho/g_c = \gamma/g$ and $V_3 = V_1$, we obtain

$$\frac{P_2 - P_3}{\gamma} = \frac{1}{g}(V_1^2 - V_1 V_2) = \frac{1}{2g}(2V_1^2 - 2V_1 V_2) \tag{2.36}$$

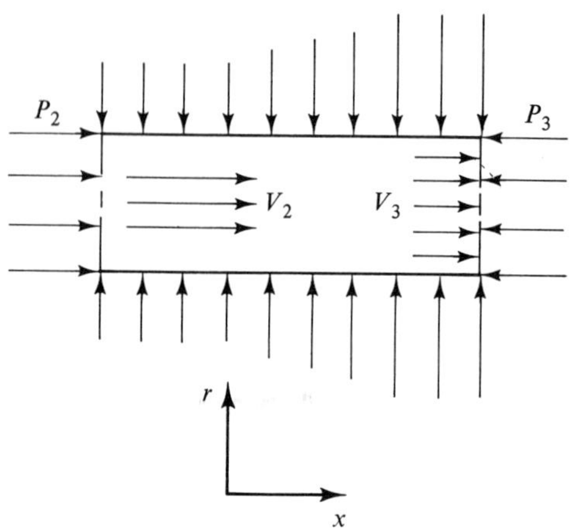

Figure 2.19 Force–momentum analysis

Adding equations (2.34) and (2.36) gives the overall head loss:

$$\begin{aligned}\frac{P_1 - P_3}{\gamma} &= \frac{P_1 - P_2}{\gamma} + \frac{P_2 - P_3}{\gamma} \\ &= \frac{1}{2g}[(V_2^2 - V_1^2) + (2V_1^2 - 2V_1V_2)] \\ &= \frac{1}{2g}(V_2^2 - 2V_1V_2 + V_1^2) \\ &= \frac{1}{2g}(V_2 - V_1)^2\end{aligned} \tag{2.37}$$

It is interesting to compare the overall pressure loss with the desired nozzle pressure drop $P_1 - P_2$ by which the flow rate is measured. Divide equation (2.37) by (2.34):

$$\frac{P_1 - P_3}{P_1 - P_2} = \frac{(V_2 - V_1)^2}{V_2^2 - V_1^2} = \frac{V_2 - V_1}{V_2 + V_1} = \frac{1 - \beta^2}{1 + \beta^2} \tag{2.38}$$

As an example, consider a nozzle with $\beta = 0.5$:

$$P_1 - P_3 = \frac{1 - \beta^2}{1 + \beta^2}(P_1 - P_2) = 0.6(P_1 - P_2)$$

For this nozzle, the overall pressure drop caused by the presence of the flow-measuring nozzle will be 60% of the pressure drop used to indicate the mass flow.

By the use of a venturi, which is much more expensive to install in a large pipe, the overall head loss can be greatly reduced. In Figure 2.13, we can apply the Bernoulli equation from point 1 to 2 to 3.

$$\frac{P_1}{\gamma} + \frac{V_1^2}{2g} = \frac{P_2}{\gamma} + \frac{V_2^2}{2g} = \frac{P_3}{\gamma} + \frac{V_3^2}{2g} \tag{2.39}$$

Since $V_3 = V_1$, we have in theory $P_3 = P_1$ with no overall pressure loss. In fact, there is some loss, particularly in the diffuser section, but in a well-designed venturi the loss can be limited to about 15% of the pressure difference $P_1 - P_2$.

REFERENCES

1. *Supplement to American Society of Mechanical Engineers Power Test Codes*, Chapter 4, "Flow Measurement," Part 5, "Measurement of Quantity of Materials" (New York: 1959).

PROBLEMS

2.1. Water at 20°C flows through a 15-cm i.d. pipe. What is the maximum flow rate if the flow is to be laminar? *Answer:* 0.00024 m³/s

2.2. Fuel oil at 50°F is to flow through a tube with $\frac{1}{4}$-in. i.d. What is the maximum flow through the tube if there is to be no turbulence in the flow? For fuel oil, $\nu = 2 \times 10^{-2}$ ft²/sec. *Answer:* 0.7 ft³/sec

2.3. Kerosene ($\nu = 2.4 \times 10^{-5}$ ft²/sec) at 70°F flows through a 1-in. pipe (1.049 in. i.d.) at the rate of 100 Imperial gpm (120 U.S. gpm). Is there likely to be any turbulence in the flow? *Answer:* Re $= 1.62 \times 10^5$

2.4. **(a)** A 3-cm-diameter nozzle is attached to a 7-cm-diameter hose. Water at 20°C flows through the hose with a velocity of 8 m/s. What is the discharge velocity? *Answer:* 44 m/s

(b) The nozzle discharges water to the atmosphere where the pressure is 101 kPa. What is the water pressure in the hose near the nozzle? *Answer:* 1018 kPa

2.5. What is the tension in the hose near the nozzle of Problem 2.4? *Answer:* 2435 N

2.6. A long horizontal tube has a bell-shaped entrance in order to bring a flow of oil in with a uniform velocity across the tube diameter. At some point downstream, laminar flow is fully developed, with the velocity profile represented by the following equation:

$$v = 3\left(1 - \frac{r^2}{R^2}\right)$$

where v is the velocity in ft/sec at the radius r. R is the radius of the tube. What is the inlet velocity of the oil? *Answer:* $V_1 = \frac{3}{2}$ ft/sec

2.7. The velocity distribution for turbulent flow in a circular pipe is well described by the expression $V_x/V_{CL} = (y/R)^{1/7}$, where R is the pipe radius, y is the distance measured from the pipe wall toward the pipe centerline, V_x is the velocity at y, and V_{CL} is the fluid velocity at the pipe centerline. At what radial location should we place a Pitot-static tube so that it will indicate the average velocity? *Answer:* $r = 0.76R$

2.8. A careless hunter with a high-powered rifle punctures a gasoline storage tank. Estimate the rate at which the fuel leaks from the tank if the jet diameter is $\frac{1}{4}$ in. and the surface of the gasoline is 30 ft above the hole. The top of the tank is at atmospheric pressure. *Answer:* 0.015 ft³/sec

2.9. A 20-cm-diameter pipe contains a short section in which the diameter is gradually reduced to 8 cm and then gradually enlarged to full size. If the pressure of water (20°C) passing through it is 200 kPa at a point just before the reduction begins, what will it be at the 8-cm section when the flow is 0.04 m³/s. Assume no loss in head occurs between the two sections. *Answer:* 169.2 kPa

2.10. Water flows from the tank through the exit pipe shown in the sketch. If d_2 = 10 cm and d_1 = 7 cm and the flow is frictionless, calculate the discharge flow rate and the absolute pressure at point 1 expressed in atmospheres. *Answer:* 0.0492 m³/s; 0.386 atmos

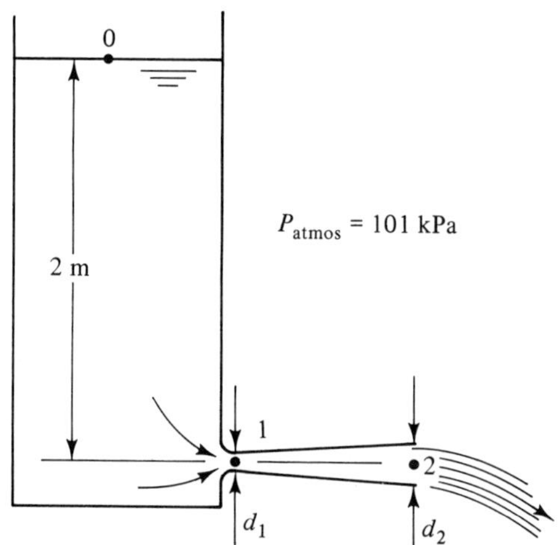

2.11. **(a)** If d_2 = 10 cm, plot a graph of P_1 absolute (expressed in atmospheres) against d_1 for values of d_1 between 5 and 10 cm.

(b) A negative absolute pressure cannot be maintained in a liquid. Vapor bubbles will form (called cavitation) when the absolute pressure is near zero. Using this hint, estimate the expected flow rate if d_1 = 5 cm. *Answer:* 0.0305 m³/s.

2.12. A 3-in. pipe discharges into the air at a point 6 ft above the level ground, the water leaving the pipe with a velocity of 28 ft/sec. Calculate the velocity of the water and the angle with which it strikes the ground (a) if the pipe is horizontal, and (b) if the pipe is inclined upward 45° from the horizontal. Assume air friction is so small as to be negligible. *Answer:* (a) θ = 35.1°, 34.21 ft/sec; (b) V_2 = 34.21 ft/sec, 54.6°

2.13. A 30-cm pipe, carrying hot oil (ρ = 812 kg/m^3) moving with an average velocity of 2 m/s, abruptly decreases in diameter to 15 cm. Compute the resultant force causing the accompanying increase in velocity. *Answer:* 689 N

2.14. A horizontal 30-cm pipe changes direction by 45°. If the velocity of the water in the pipe is 3 m/s, what resultant force components, ΣF_x and ΣF_y, will be exerted on the water in the bend? Take the x and y axes to be parallel and normal, respectively, to the original direction of the pipe. *Answer:* −186 N; +449 N

2.15. Water flows radially outward in all directions from between two horizontal circular plates that are parallel, 4 ft in diameter, and separated by a distance of 1 in. A flow of 1 ft^3/sec is maintained by a pipe entering one of the plates at its center. What pressure will exist between the plates at a point 6 in. from the center? Assume no loss of head occurs. *Answer:* −0.092 psig

2.16. Two vessels, A and B, containing water under pressure, are connected by an oil differential manometer as illustrated. Find the difference in pressure between the vessels A and B. Specific gravity of the oil is 0.80. *Answer:* 1.99 psi

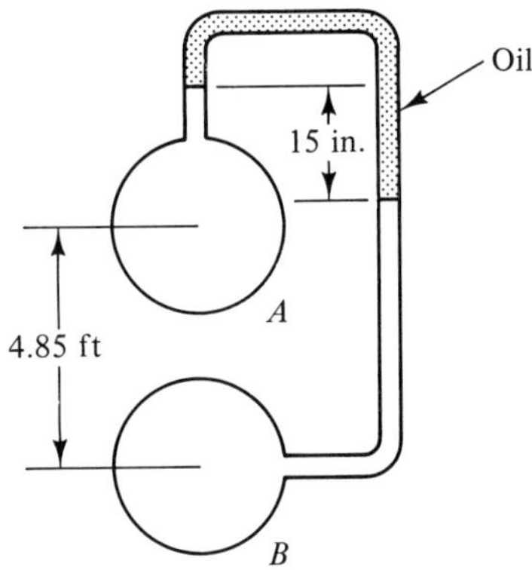

2.17. A liquid (SG 1.25) partly fills the vessel A as illustrated. What is the pressure at a point 0.6 m below its surface if the height of the mercury column is 50 cm? *Answer:* −59.2 kPa gauge

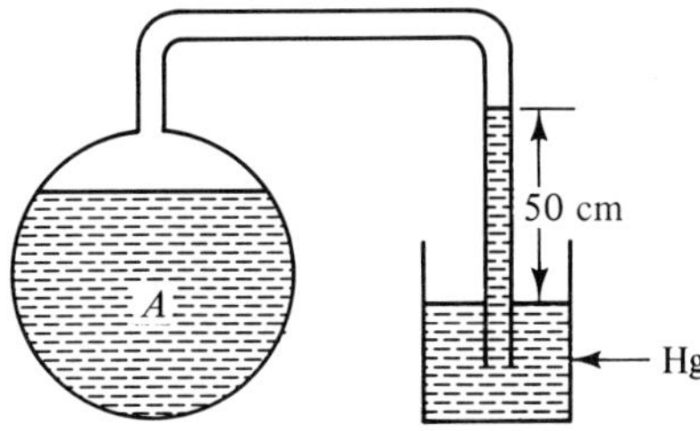

2.18. The inclined manometer illustrated is connected at A to a small experimental wind tunnel. What is the pressure at A if the reading on the manometer is 20

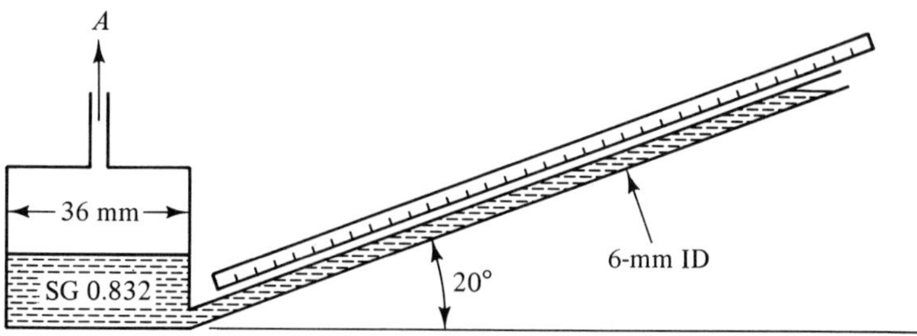

cm? The liquid meniscus in the sloping tube is at zero on the scale when the reservoir is open to atmosphere. If the scale can be read to the nearest millimeter, is it necessary to correct for the change in fluid level in the reservoir? *Answer:* The *correction* for change of reservoir level *is necessary;* 602.4 N/m^2

2.19. A Pitot-static tube placed at the center of a 30-cm pipe carrying air at 100 kPa and 25°C registered a differential pressure head of 30 cm of water. Calculate the air velocity at the point of measurement. Can this velocity be used to calculate the rate of flow through the pipe? Explain. *Answer:* v = 70.9 m/s

2.20. Air at a temperature of 60°F flows in a 6-in. pipe (6.065 in. i.d.), the static pressure at a certain section being 24.7 psia. A Pitot-static tube at the center of the same section produces a deflection of 12 in. of water in a differential U-tube. Estimate the rate of flow of the air in lb_m/sec. What important assumption is involved in this estimate? *Answer:* 4.55 lb_m/sec

2.21. A venturi meter has an area ratio of 9 to 1, the larger diameter being 30 cm. Installed in a line carrying water at 20°C, the gauge pressure in the large section is 6.51 m of water, while that at the throat is 4.24 m of water. What is the volume flow rate through the meter? *Answer:* 0.0520 m^3/s

2.22. A 5-in.-diameter venturi meter has a throat diameter of 3 in. During a 10-minute test it discharged 1160 ft^3 of water when the measured upstream gauge pressure was 9.70 ft of water and the throat pressure was 10.52 in. of mercury vacuum. Calculate the meter discharge coefficient and compare with the value obtained from Figure 2.15. *Answer:* 0.985

2.23. The discharge through a flow nozzle is 920 Imperial gallons of water per minute. The diameter of the pipe is 10 in. and the pipe-to-throat area ratio is 4 to 1. If the pressure at the entrance is 18.8 ft of water, find the average velocity and the pressure at the throat of the nozzle (use Figure 2.15). *Answer:* 18.0 ft/sec; 14.0 ft of water

2.24. A sharp-edged circular orifice is to be made to measure water flowing at a rate not to exceed 34 m^3/hr, with a differential head of 3 m. What orifice diameter is required if D_1/D_o is made 5 to 1? *Answer:* 5.07 cm

2.25. A standard sharp-edged orifice 2.5 cm in diameter is to be installed in a 7.5-cm-diameter pipeline. Compressed air at 780 kPa and 40°C flows through the meter. When the reading of the manometer across the meter is 36 cm of Hg, what is the flow rate of the air in kg/s? *Answer:* 0.275 kg/s

2.26. Crude oil (SG 0.87) is flowing at 150°F through a 6-in. pipe (6.065 in. i.d.). The maximum flow rate through the line is 1000 bbl/hr (1 bbl ≡ 35 Imperial gal = 42 U.S. gals). To measure the flow, a standard sharp-edged orifice with flange taps is to be installed in the line. The pressure difference measured at the flange taps is not to exceed 3.6 psi. Specify the orifice diameter. *Answer:* Required orifice diameter = 4.10 in.

2.27. For Problems 2.25 and 2.26, determine the power loss due to the presence of the meter in the line. For your analysis, assume that the fluid velocity at the vena contracta is uniform and may be derived by application of Bernoulli's equation across the orifice. *Answer:* Problem 25, 1327 W; Problem 26, 0.79 hp

Applications in One-Dimensional Fluid Flow

3

3.1 Effect of Friction on the Pressure–Velocity Relation

It is intuitively obvious that energy dissipation by friction in pipe flow will cause a pressure reduction in the flow direction. While Bernoulli's equation is valid only for frictionless flow, it can be modified to take account of frictional effects. With frictional effects included,

$$\frac{P_1}{\gamma} + \frac{V_1^2}{2g} + Z_1 > \frac{P_2}{\gamma} + \frac{V_2^2}{2g} + Z_2 \tag{3.1}$$

This intuitively obvious statement can be expressed as an equality by addition of a *head-loss* term.

$$\frac{P_1}{\gamma} + \frac{V_1^2}{2g} + Z_1 = \frac{P_2}{\gamma} + \frac{V_2^2}{2g} + Z_2 + h_f \tag{3.2}$$

where h_f is the head loss due to frictional effects between station 1 and station 2.

3.2 Steady-Flow Energy Equation

The pressure–velocity relation, including friction effects, can also be derived from a consideration of the principle of energy conservation (Figure 3.1).

h	enthalpy per unit mass
ke	kinetic energy per unit mass
pe	potential energy per unit mass
δq	heat added per unit mass
δw	work extracted per unit mass
δlw	lost work due to friction per unit mass
s	entropy per unit mass

For steady flow, the energy entering the system equals the energy leaving.

$$\delta q = dh + d\,ke + d\,pe + \delta w \tag{3.3}$$

Some thermodynamic manipulations can be used to transform equation (3.3) into a form rather similar to Bernoulli's equation.

$T\,ds = \delta q + \delta lw,$ the entropy of a system is increased by heat addition and/or mechanical energy dissipation

and

$T\,ds = dh - \dfrac{dP}{\rho},$ a property relation derived from the first and second laws of thermodynamics

Thus

$$\delta q = dh - \frac{dP}{\rho} - \delta lw$$

Substituting this expression for δq into equation (3.3),

$$dh - \frac{dP}{\rho} - \delta lw = dh + d\,ke + d\,pe + \delta w$$

Canceling dh on each side and substituting

$$d\,ke = \frac{V\,dV}{g_c} \quad \text{and} \quad d\,pe = \frac{g}{g_c}\,d\,Z$$

Figure 3.1 Energy flows

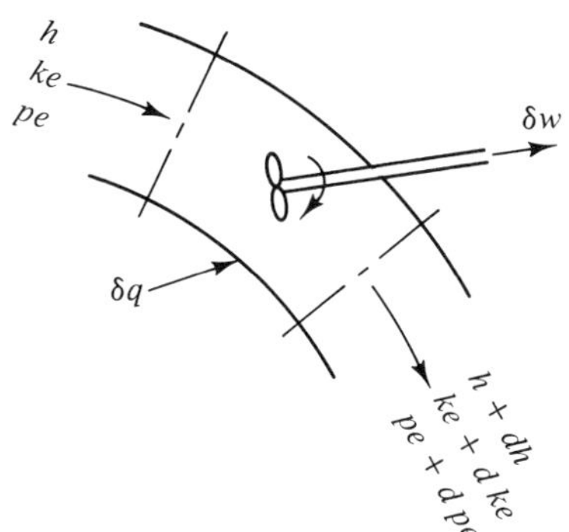

we get the differential form of the mechanical energy equation:

$$\frac{dp}{\rho} + \frac{V\,dV}{g_c} + \frac{g}{g_c}\,d\,Z + \delta lw + \delta w = 0 \tag{3.4}$$

Integrating from station 1 to station 2, for incompressible flow

$$\frac{P_1}{\rho} + \frac{V_1^2}{2g_c} + \frac{g}{g_c} Z_1 = \frac{P_2}{\rho} + \frac{V_2^2}{2g_c} + \frac{g}{g_c} Z_2 + lw + w_{out} \tag{3.5}$$

Multiply by g_c/g to express the energy equation in terms of head.

$$\frac{P_1}{\gamma} + \frac{V_1^2}{2g} + Z_1 = \frac{P_2}{\gamma} + \frac{V_2^2}{2g} + Z_2 + h_f + w_{out} \tag{3.6}$$

where the work output is now in energy units per unit weight, and h_f is the head loss due to friction, $(g_c/g) \times lw$. If there is work input from a pump, rather than work output (as from a turbine), then the energy equation would be expressed as follows:

$$\frac{P_1}{\gamma} + \frac{V_1^2}{2g} + Z_1 + w_{in} = \frac{P_2}{\gamma} + \frac{V_2^2}{2g} + Z_2 + h_f \tag{3.7}$$

3.3 Kinetic Energy Correction Factor α

If the fluid velocity at a stream section is not uniform, then the kinetic energy is not properly evaluated by using the average velocity in the expression $V^2/2g$. The average kinetic energy per unit weight of fluid is expressed as $\alpha\overline{V}^2/2g$, where the correction factor α depends on the shape of the velocity distribution (Figure 3.2).

$$d\,\dot{KE} = \gamma\,V\,dA \times V^2/2g$$

$$\dot{KE} = \frac{\gamma}{2g}\int_A V^3\,dA = \gamma A\overline{V} \times \alpha\overline{V}^2/2g \tag{3.8}$$

$$\alpha = \frac{(\gamma/2g) \times \int_A V^3\,dA}{(\gamma/2g) \times A\overline{V}^3} = \frac{1}{A}\int_A \left(\frac{V}{\overline{V}}\right)^3 dA$$

With the kinetic energy correction factor included, the energy equation takes the form

$$\frac{P_1}{\gamma} + \frac{\alpha_1\overline{V}_1^2}{2g} + Z_1 + w_{in} = \frac{P_2}{\gamma} + \frac{\alpha_2\overline{V}_2^2}{2g} + Z_2 + h_f \tag{3.9}$$

where $\overline{V}$ is the average velocity at the stream section.

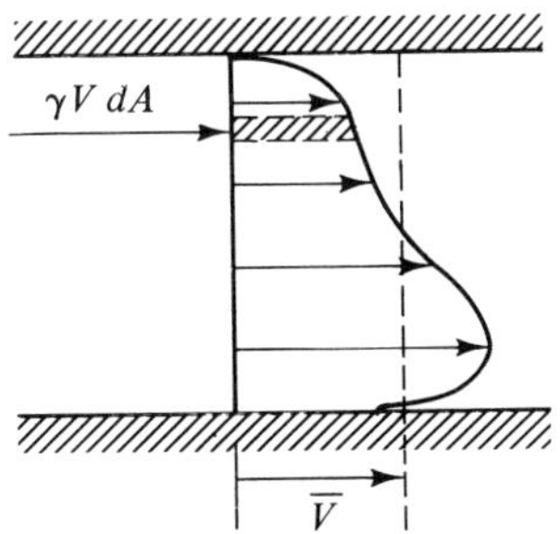

Figure 3.2 Velocity distribution

3.4 Dimensional Analysis Applied to Pipe Friction

The overall head loss caused by the presence of a nozzle in a pipe, dealt with in Section 2.7.5, is one case that can be handled analytically. In general, however, head losses in various applications must be measured experimentally. When the number of variables is large, dimensional analysis is used to limit the number of measurements required and to allow a coherent presentation of data.

Consider the case of pressure loss per unit length of pipe, $\Delta P/L$, caused by friction. It should be apparent that the pressure drop per unit length will be affected by changes in the following quantities: fluid density, velocity and viscosity, and pipe diameter and surface roughness. Acknowledging some uncertainty about the completeness of the list of variables, we can tentatively express the following relationship:

$$\frac{\Delta P}{L} = f(\rho, V, \mu, D, \varepsilon) \tag{3.10}$$

where ε is the roughness expressed in the same units as the diameter D. How do we find the nature of this complicated relationship experimentally?

Let us digress momentarily to consider a simpler problem. Suppose we want to determine the elevation of a lawn as a function of position.

$$z = f(x, y)$$

We could lay out a grid of ten units in the x direction and ten units in the y direction and measure the elevations with a surveyor's level at the grid intersections ($10 \times 10 = 100$ measurements of elevation). We could present the information as a contour map or in the form of a table giving values of z at all possible combinations of x and y.

Returning to our pressure-loss problem, if we set out to determine the nature of the relationship in the same direct manner as for the lawn elevations, with ten values of each variable, we would have 100,000 experiments to perform. Even if we could do it, how would we present the information? Dimensional analysis is the key to this problem.

Equations describing physical phenomena must be dimensionally homogeneous. In our case, the dimensions of the function $f(\rho, V, \mu, D, \varepsilon)$ must be the same as the dimensions of $\Delta P/L$. (For those who need a reminder, *units* are lb_f, lb_m, ft, sec, newtons, kilograms, meters, etc., and *dimensions* are the basic quantities force, mass, length, and time.) The dimensions of $\Delta P/L$ are F/L^3, and only certain groupings of ρ, V, μ, D, ε can produce these same dimensions. Arranging the quantities into these groups in effect reduces the number of independent variables. In fact, if there are n quantities (ρ, V, μ, D, ε) and m dimensions represented among them, the number of independent variables (groups) will be $n - m$.

Before embarking on the procedure (dimensional analysis) to arrange our quantities (ρ, V, μ, D, ε) in groups, some further clarification of units and dimensional systems is required. The commonly used procedure is to use mass, length, and time as the basic dimensions, and the dimensions of force are derived using Newton's second law, $F = ma$. Thus the dimensions of force would be MLt^{-2}. This method is used with the SI system.

In the English engineering system, four basic dimensions are used, *force*, *mass*, *length*, and *time*, with Newton's second law expressed as $F = (m/g_c)\,a$. When using this system in dimensional analysis, we must add g_c to the list of quantities to provide the necessary connection between the force dimension and the other three of mass, length, and time.

We are now ready to condense our list of variables into groups by the use of dimensional analysis. Using the English engineering system, we begin with the statement

$$\frac{\Delta P}{L} = f(\rho, V, \mu, D, \varepsilon, g_c) \tag{3.11}$$

Let us assume that this functional relation can be expressed as an infinite power series:

$$\begin{aligned} \frac{\Delta P}{L} &= C_1(\rho^{a1}\, V^{b1}\, \mu^{c1}\, D^{d1}\, \varepsilon^{e1}\, g_c^{f1}) \\ &\quad + C_2(\rho^{a2}\, V^{b2}\, \mu^{c2}\, D^{d2}\, \varepsilon^{e2}\, g_c^{f2}) + C_3\,(\ldots) \end{aligned} \tag{3.12}$$

where $C_1, C_2, C_3, \ldots, C_n$ are dimensionless constants.
Let us take just one of the terms in the series and drop the numerical subscript for convenience:

$$\frac{\Delta P}{L} \overset{\text{D}}{=} (\rho^{a}\, V^{b}\, \mu^{c}\, D^{d}\, \varepsilon^{e}\, g_c^{f}) \tag{3.13}$$

where the sign $\overset{\text{D}}{=}$ means *dimensionally equal*. In the F, M, L, t system, the dimensions of the various quantities are as given in Table 3.1.

TABLE 3.1

Quantity		Units	Dimensions
$\dfrac{\Delta P}{L}$		lb_f/ft^3	$F\,L^{-3}$
ρ		lb_m/ft^3	$M\,L^{-3}$
V		ft/sec	$L\,t^{-1}$
μ	$\dfrac{F/A}{\partial V/\partial y}$	$lb_f\ sec/ft^2$	$F\,t\,L^{-2}$
D		ft	L
ε		ft	L
g_c		$\dfrac{lb_m\ ft}{lb_f\ sec^2}$	$M\,L\,F^{-1}\,t^{-2}$

Referring to equation (3.13), we can balance the dimensions of mass on either side of the equation and then do the same thing for force, length, and time.

	Left side	$\stackrel{D}{=}$	Right side	
M	0	$=$	$a + f$	(3.14)
F	1	$=$	$c - f$	(3.15)
L	-3	$=$	$-3a + b - 2c + d + e + f$	(3.16)
t	0	$=$	$-b + c - 2f$	(3.17)

With six unknowns (a, b, c, d, e, and f) and four equations, we should be able to write four of the unknowns in terms of the remaining two. Let us solve for a, b, d, and f in terms of c and e. (There is a little more to this choice than meets the eye.)

From (3.15) $\quad f = c - 1$

From (3.14) $\quad a = -f = -c + 1$

From (3.17) $\quad b = c - 2f = c - 2c + 2 = -c + 2$

From (3.16)

$$
\begin{aligned}
d &= -3 + 3a - b + 2c - e - f \\
&= -3 + 3(1 - c) - (2 - c) + 2c - e - (c - 1) \\
&= -1 - c - e
\end{aligned}
$$

Substitute back in equation (3.13):

$$
\begin{aligned}
\frac{\Delta P}{L} &\stackrel{D}{=} (\rho^{-c+1} V^{-c+2} \mu^c D^{-1-c-e} \varepsilon^e g_c{}^{c-1} \\
&\stackrel{D}{=} \frac{\rho V^2}{D g_c}\left[\left(\frac{\mu g_c}{\rho V D}\right)^c \left(\frac{\varepsilon}{D}\right)^e\right]
\end{aligned}
\tag{3.18}
$$

Now return to equation (3.12), remembering that (3.18) is the result of dealing with one sample term in (3.12).

$$
\begin{aligned}
\frac{\Delta P}{L} &= C_1\frac{\rho V^2}{Dg_c}\left\{\left[\left(\frac{\mu g_c}{\rho VD}\right)^{c_1}\left(\frac{\varepsilon}{D}\right)^{e_1}\right]\right. \\
&\quad \left. + C_2\frac{\rho V^2}{Dg_c}\left[\left(\frac{\mu g_c}{\rho VD}\right)^{c_2}\left(\frac{\varepsilon}{D}\right)^{e_2}\right] + C_3 \ldots\right\} \\
&= \frac{\rho V^2}{Dg_c}\left\{C_1\left[\left(\frac{\mu g_c}{\rho VD}\right)^{c_1}\left(\frac{\varepsilon}{D}\right)^{e_1}\right]\right. \qquad (3.19) \\
&\quad \left. + C_2\left[\left(\frac{\mu g_c}{\rho VD}\right)^{c_2}\left(\frac{\varepsilon}{D}\right)^{e_2}\right] + C_3 \ldots\right\} \\
&= \frac{\rho V^2}{Dg_c} f\left(\frac{\mu g_c}{\rho VD}, \frac{\varepsilon}{D}\right)
\end{aligned}
$$

Some observations regarding equation (3.19) are in order.

1. The terms $\mu g_c/\rho VD$ and ε/D are both dimensionless. The group $\mu g_c/\rho VD$ is the Reynold's number inverted. If something is a function of 1/Re, it is also a function of Re.
2. We can divide through equation (3.19) by $\rho V^2/Dg_c$:

$$\frac{\Delta P/L}{\rho V^2/Dg_c} = f^1\left(\text{Re}, \frac{\varepsilon}{D}\right)$$

Note that $\rho/g_c = \gamma/g$, so we can rearrange the appearance of the left side.

$$\frac{\Delta P/\gamma}{(L/D)(V^2/g)} = f^1\left(\text{Re}, \frac{\varepsilon}{D}\right)$$

3. $\Delta P/\gamma$ is equal to h_f, the head loss due to friction. Also, if we multiply the V^2/g by $\frac{1}{2}$, it will represent the kinetic energy.

$$\frac{h_f}{(L/D)(V^2/2g)} = 2f^1\left(\text{Re}, \frac{\varepsilon}{D}\right) = f^{11}\left(\text{Re}, \frac{\varepsilon}{D}\right)$$

4. The quantity $h_f/(L/D)(V^2/2g)$ is dimensionless and is called the friction factor f.

$$f = \frac{h_f}{(L/D)(V^2/2g)} = \text{a function of } \left(\text{Re}, \frac{\varepsilon}{D}\right) \qquad (3.20)$$

5. The friction factor f for pipe flow has been determined experimentally as a function of Reynold's number Re and the relative roughness ε/D, and the information is presented in the Moody diagram of Figure 3.3.

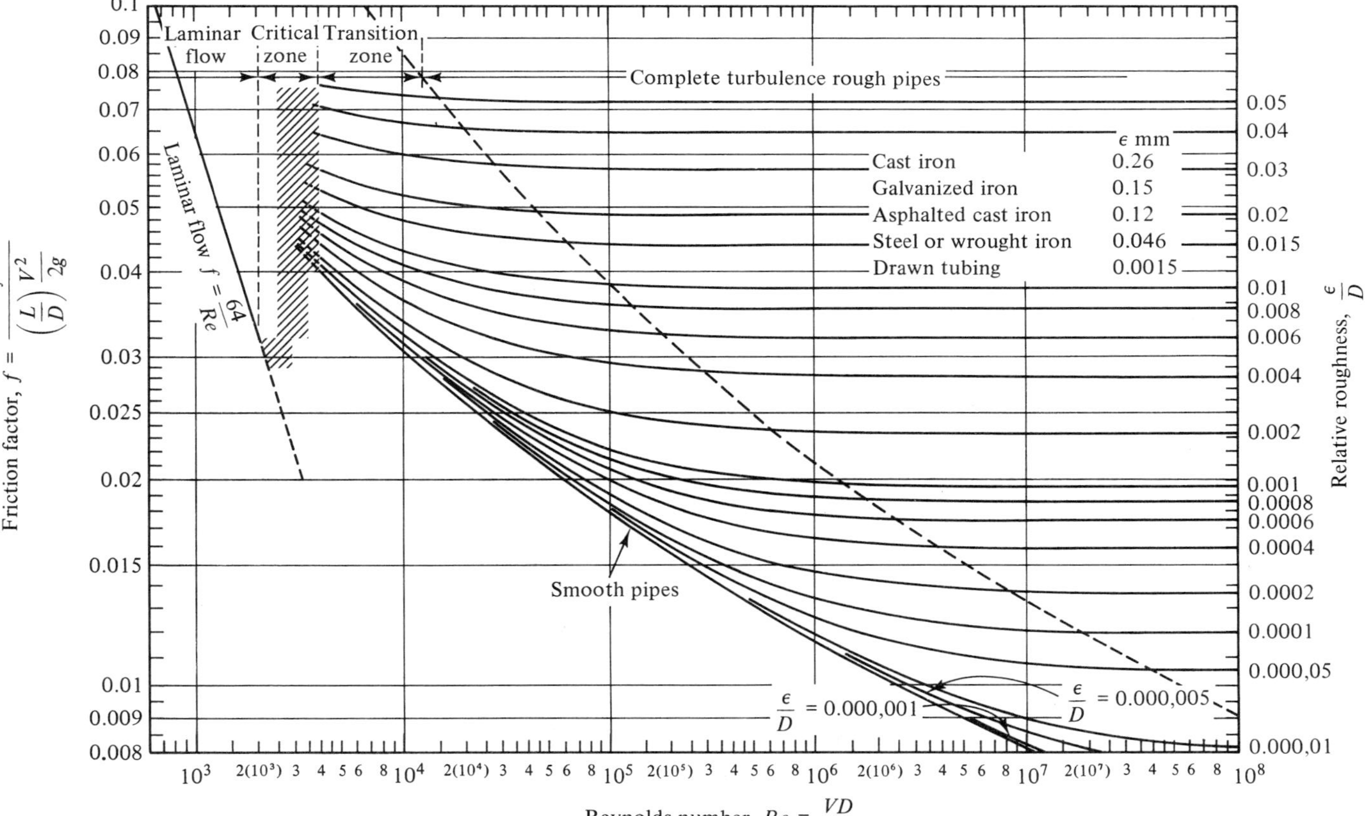

Figure 3.3 Moody diagram (adapted from Moody, L.F., *Trans. ASME, 66*, 1944, p. 672)

6. When the friction factor f has been read from the Moody diagram, the head loss due to friction can be calculated from equation (3.20).

$$h_f = f\frac{L}{D}\frac{V^2}{2g} \tag{3.21}$$

3.5 Problem Types Using the Friction Factor

Because the friction factor is expressed graphically rather than as a mathematical equation, problem solutions sometimes require a trial-and-error approach. The following examples will illustrate various problem types.

3.5.1 Velocity Known, Find Pressure Loss: Direct Solution. Water at 20°C flows through a 15-cm-diameter cast-iron pipe at the rate of 5 m^3/min. What is the pressure loss due to friction per 100 m of pipe? The viscosity of water at 20°C is $\mu = 0.00101$ kg/m-s. The surface roughness of cast iron is $\varepsilon = 0.26$ mm. The velocity of water flowing in pipe is

$$V = \frac{Q}{A} = \frac{5/60\ (\text{m}^3/\text{s})}{(\pi/4)(0.15)^2\ \text{m}^2} = 4.72\ \text{m/s}$$

The Reynold's number is

$$\text{Re} = \frac{\rho VD}{\mu} = \frac{998\ (\text{kg/m}^3) \times 4.72\ (\text{m/s}) \times 0.15\ \text{m}}{0.00101\ \text{kg/m-s}} = 700{,}000$$

The relative roughness is

$$\frac{\varepsilon}{D} = \frac{0.026\ \text{cm}}{15\ \text{cm}} = 0.0017$$

Using the Moody diagram, find that $f = 0.0225$. Thus

$$h_f = f\frac{L}{D}\frac{V^2}{2g} = 0.0225 \times \frac{100}{0.15} \times \frac{(4.72)^2}{2 \times 9.81} = 17.0\ \text{m}$$

The pressure drop due to friction is

$$\begin{aligned} P_1 - P_2 = \gamma h_f &= \rho g h_f \\ &= 998\ \text{kg/m}^3 \times 9.81\ \text{m/s}^2 \times 17.0\ \text{m} \\ &= 167{,}000\ \text{kg/m s}^2 \\ &= 167{,}000\ (\text{kg m/s}^2) \times (1/\text{m}^2) \quad \text{or} \quad \text{N/m}^2 \\ &= 167\ \text{kPa per 100 m of pipe} \end{aligned}$$

3.5.2 Pressure Loss Known, Find Velocity: Trial and Error. Find the volume flow rate of water through a 2-in.-diameter commercial steel pipe if the pressure loss per foot is 0.1 psi.

$$\text{Viscosity:} \quad \mu = 2.36 \times 10^{-5}\ \text{lb}_f\text{-sec/ft}^2$$

$$\text{Relative roughness:} \quad \frac{\varepsilon}{D} = \frac{0.046}{2 \times 25.4} = 0.0009$$

Using the Moody diagram at $\varepsilon/D = 0.0009$ (Re unknown).

$$\text{Estimate:} \quad f = 0.0195$$

Now

$$h_f = f\frac{L}{D}\frac{V^2}{2g} = \frac{\Delta P}{\gamma} = \frac{0.1 \times 144}{62.4} = 0.231 \text{ ft per ft of pipe}$$

and

$$V = \sqrt{\frac{2gh_f}{f\,(L/D)}} = \sqrt{\frac{64.4 \times 0.231}{0.0195 \times [1/(2/12)]}} = 11.28 \text{ ft/sec}$$

$$\text{Re} = \frac{\rho VD}{\mu} = \frac{62.4 \times 11.28 \times (2/12)}{2.36 \times 10^{-5} \times 32.2} = 154{,}300$$

Now check the friction factor for $\varepsilon/D = 0.0009$ and $Re = 154{,}300$. From the Moody diagram, $f = 0.021$. Since our first estimate of f ($= 0.0195$) was not close enough, we must repeat the calculation of velocity.

$$V = \sqrt{\frac{2gh_f}{f(L/D)}} = \sqrt{\frac{64.4 \times 0.231}{0.021 \times [1/(2/12)]}} = 10.87 \text{ ft/sec}$$

$$Re = \frac{62.4 \times 10.87 \times 2/12}{2.36 \times 10^{-5} \times 32.2} = 148{,}800$$

Check f again. From the Moody diagram, $f \simeq 0.021$. Thus, the volume flow rate through the pipe is

$$Q = AV = \frac{\pi}{4}\left(\frac{2}{12}\right)^2 \times 10.87 = 0.237 \text{ ft}^3\text{/sec}$$

3.5.3 Head Loss and Flow Rate Known: Choose Pipe Size. What size of asphalted cast-iron pipe is required to carry 1.7 million cubic meters of water per day with a head loss of 20 m/km of pipe?

Viscosity of water at 20°C: $\mu = 0.00101$ kg/s-m

Surface roughness: $\varepsilon = 0.00012$ m

$$\text{Head loss is } h_f = f\frac{L}{D}\frac{V^2}{2g} = 20 \text{ m/km of pipe}$$

$$\text{Flow rate: } Q = \frac{\pi D^2}{4} \times V = \frac{1.7 \times 10^6}{24 \times 60 \times 60} = 19.7 \text{ m}^3/\text{s}$$

$$\text{Velocity: } V = \frac{19.7}{\pi D^2/4} = \frac{25.1}{D^2} \text{ m/s}$$

Substitute in the expression for head loss:

$$h_f = f\frac{L}{D}\left(\frac{25.1}{D^2}\right)^2 \frac{1}{2g} = 20$$

$$f = \frac{20 \times 2g \times D^5}{1000 \times (25.1)^2} = 0.000623D^5$$

Now we could make a guess about either f or D. Because D is raised to the fifth power, faster convergence of the trial solution will be achieved by guessing f.

Guess $f = 0.02$. Then

$$D = \left(\frac{0.02}{0.000623}\right)^{1/5} = 2.0 \text{ m}$$

and

$$V = \frac{25.1}{D^2} = 6.28 \text{ m/s}$$

$$\text{Re} = \frac{\rho V D}{\mu} = \frac{998 \times 6.28 \times 2.0}{0.00101} = 12.4 \times 10^6$$

$$\frac{\varepsilon}{D} = \frac{0.00012}{2.0} = 0.00006$$

Using the Moody diagram, with $\text{Re} = 12.4 \times 10^6$ and $\varepsilon/D = 0.00006$, the next estimate for the friction factor is

$$f = 0.0126$$

We now repeat the calculations for this new value of f.

$$D = \left(\frac{0.0126}{0.000623}\right)^{1/5} = 1.82 \text{ m}$$

$$V = \frac{25.1}{D^2} = 7.58 \text{ m/s}$$

$$\text{Re} = \frac{998 \times 7.58 \times 1.82}{0.00101} = 13.6 \times 10^6$$

$$\frac{\varepsilon}{D} = \frac{0.00012}{1.82} = 0.000066$$

$$f = 0.0127$$

This is close enough to the previous value of f to consider that we have a valid solution. Therefore, the required pipe size is 1.82 m in diameter.

3.6 Laminar Flow in Pipes

Laminar flow in a pipe is another example that can be handled analytically. While laminar-flow problems can be solved using the Moody diagram, we get more insight into laminar-flow behavior from an analytical treatment.

Consider a force balance on a cylindrical plug of fluid (Figure 3.4) undergoing laminar flow in a pipe. For steady conditions, the resultant force acting on the plug must be zero by Newton's second law.

$$\sum F_x = 0 \tag{3.22}$$

$$(P_1 - P_2)\pi r^2 + 2\pi r L \tau = 0$$

Newton's law of viscosity is

$$\tau = \frac{F}{A} = \mu \frac{dV}{dr} \tag{3.23}$$

Substitute equation (3.23) into (3.22), rearrange and integrate:

$$dV = -\frac{P_1 - P_2}{2\mu L} r\, dr$$

$$V = -\frac{P_1 - P_2}{4\mu L} r^2 + C$$

Figure 3.4 Laminar flow in a pipe

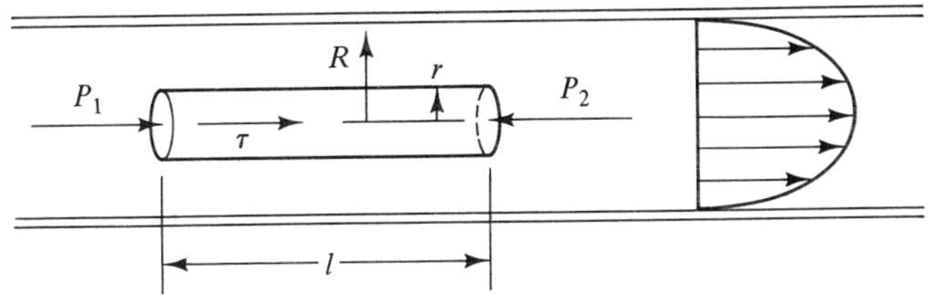

At $r = R$, $V = 0$ because there is no slip at the wall. Therefore,

$$C = \frac{P_1 - P_2}{4\mu L} R^2$$
$$V = \frac{P_1 - P_2}{4\mu L} (R^2 - r^2) \tag{3.24}$$

At $r = 0$, the pipe centerline, we have

$$V = \frac{P_1 - P_2}{4\mu L} R^2 = V_{\max} \tag{3.25}$$

Thus the laminar velocity profile is given by

$$V = V_{\max}\left[1 - \left(\frac{r}{R}\right)^2\right] \tag{3.26}$$

which is a parabolic velocity distribution.

3.6.1 Laminar Flow Rate and Pressure Drop.

The rate of flow through the ring element dA in Figure 3.5 is

$$dQ = V\,dA = \frac{P_1 - P_2}{4\mu L} (R^2 - r^2)2\pi r\,dr$$

The total flow rate is

$$Q = \overline{V}A = \frac{P_1 - P_2}{4\mu L} \int_{r=0}^{r=R} (R^2 - r^2)2\pi r\,dr$$
$$= \frac{P_1 - P_2}{8\mu L} \pi R^4 \tag{3.27}$$
$$= \frac{P_1 - P_2}{128\mu L} \pi D^4$$

Equation (3.27) is known as the Hagen–Poiseuille equation. It can be used to find Q if the pressure drop is known, or to find the pressure loss if Q is specified.

Figure 3.5 Velocity profile for laminar flow in a tube

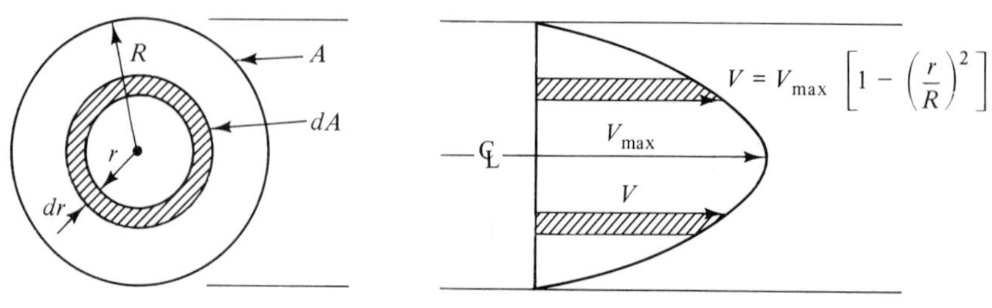

The average velocity is

$$\overline{V} = \frac{Q}{A} = \frac{(P_1 - P_2)/(8\mu L)\pi R^4}{\pi R^2} = \frac{P_1 - P_2}{8\mu L} R^2$$

Comparing with $V_{\max} = (P_1 - P_2)R^2/(4\mu L)$ [equation (3.25)], we observe that the average velocity in laminar flow is one-half the maximum centerline velocity.

$$\overline{V}_{\text{laminar}} = \tfrac{1}{2} V_{\max} \tag{3.28}$$

3.6.2 Friction Factor for Laminar Flow.

The friction factor for laminar flow can be found from the definition of friction factor and the Hagen–Poiseuille equation. The definition of f is

$$f = \frac{h_f}{(L/D)(\overline{V}^2/2g)} = \frac{(P_1 - P_2)/\gamma}{(L/D)(\overline{V}^2/2g)} \tag{3.29}$$

and from equation (3.27)

$$\frac{P_1 - P_2}{\gamma} = \frac{128\mu LQ}{\gamma\pi D^4} = \frac{128\mu LA\overline{V}}{4\gamma(\pi D^2/4)D^2} = \frac{32\mu L\overline{V}}{\gamma D^2} \tag{3.30}$$

Hence

$$f = \frac{(32\mu L\overline{V})/(\gamma D^2)}{(L/D)(\overline{V}^2/2g)} = \frac{64\mu g}{\gamma\overline{V}D} = \frac{64}{\text{Re}} \tag{3.31}$$

The relation appears as a straight line on the Moody diagram, which has log–log coordinates:

$$\log f = \log 64 - \log \text{Re}$$

3.6.3 Kinetic Energy Correction Factor in Laminar Flow.

In Section 3.3 we derived the general equation for α. We now apply equation (3.8) to the particular velocity profile for laminar flow given by equation (3.26).

$$\alpha = \frac{1}{A}\int_A \left(\frac{V}{\overline{V}}\right)^3 dA \tag{3.8}$$

$$V = V_{\max}\left[1 - \left(\frac{r}{R}\right)^2\right] = 2\overline{V}\left[1 - \left(\frac{r}{R}\right)^2\right] \tag{3.26}$$

after using (3.28). Thus,

$$\begin{aligned}\alpha &= \frac{1}{\pi R^2}\int_{r=0}^{r=R}\left\{2\left[1 - \left(\frac{r}{R}\right)^2\right]\right\}^3 2\pi r\, dr \\ &= 8\int_0^1 \left[1 - \left(\frac{r}{R}\right)^2\right]^3 2\frac{r}{R}\, d\left(\frac{r}{R}\right)\end{aligned} \tag{3.32}$$

Let $Z = 1 - (r/R)^2$; then

$$dZ = -2 \frac{r}{R} d\left(\frac{r}{R}\right)$$

At $r/R = 0$, $Z = 1$, and at $r/R = 1$, $Z = 0$. We can rewrite (3.32) in much simpler form:

$$\alpha = -8\int_1^0 Z^3\, dZ = -8 \frac{Z^4}{4}\Bigg|_1^0 = +2$$

Therefore, the kinetic energy correction factor for laminar flow is $\alpha = 2$.

3.6.4 Summary of Results for Laminar Flow. The following easily remembered facts provide a convenient basis for the solution of problems in laminar pipe flow.

$$V = \frac{P_1 - P_2}{4\mu L}(R^2 - r^2)$$

$$\overline{V} = \tfrac{1}{2} V_{max}$$

$$f = \frac{64}{\text{Re}}$$

$$\alpha = 2$$

3.7 Minor Losses in Pipe Flow

Minor head losses in addition to pipe friction losses are caused by any pipe fittings that disturb the flow pattern. Valves, elbows, changes in pipe size, and the like, all cause pressure losses that must be accounted for, at least in short pipe systems. In long pipelines, the losses due to fittings are generally considered negligible. These losses may be calculated using an experimentally determined loss coefficient multiplied by the velocity head:

$$h_f = K_L \frac{V^2}{2g} \tag{3.33}$$

or, alternatively, the effect of the fitting may be replaced by the addition of an equivalent length of pipe.

$$h_f = K_L \frac{V^2}{2g} = f \frac{L_{eq}}{D} \frac{V^2}{2g}$$

$$L_{eq} = \frac{K_L D}{f} \tag{3.34}$$

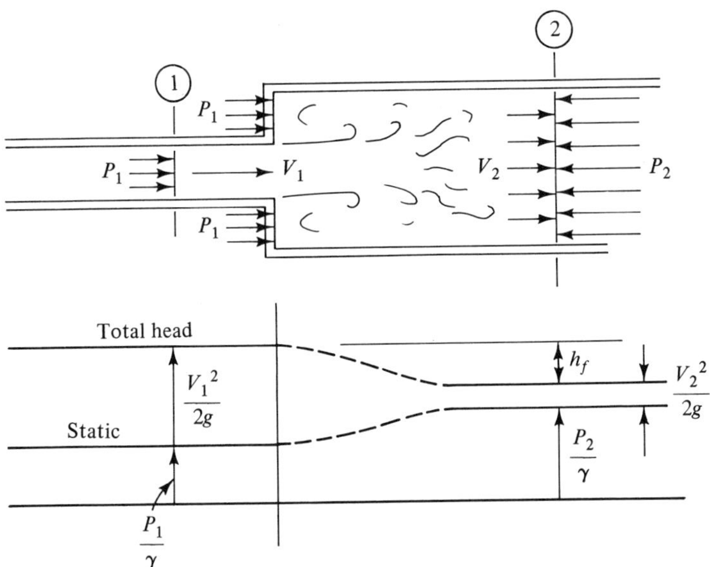

Figure 3.6 Head loss due to a sudden enlargement

3.7.1 Head Loss Due to a Sudden Enlargement.

The head loss due to a sudden enlargement is the only minor loss that can be determined purely analytically. The treatment is essentially the same as used earlier (Section 2.7.5) to determine the overall head loss due to a nozzle.

Referring to Figure 3.6, the energy equation is

$$\frac{P_1}{\gamma} + \frac{V_1^2}{2g} = \frac{P_2}{\gamma} + \frac{V_2^2}{2g} + h_f$$

from which

$$h_f = \frac{P_1 - P_2}{\gamma} + \frac{V_1^2 - V_2^2}{2g} \tag{3.35}$$

Applying the linear momentum equation between points 1 and 2 gives

$$\sum F_x = \dot{m}(V_2 - V_1)$$

$$(P_1 - P_2)A_2 = \frac{\gamma A_2 V_2}{g}(V_2 - V_1)$$

from which

$$\frac{P_1 - P_2}{\gamma} = \frac{V_2^2 - V_1 V_2}{g} = \frac{2V_2^2 - 2V_1 V_2}{2g} \tag{3.36}$$

Substitute equation (3.36) in (3.35):

$$
\begin{aligned}
h_f &= \frac{2V_2^2 - 2V_1V_2}{2g} + \frac{V_1^2 - V_2^2}{2g} \\
&= \frac{V_1^2 - 2V_1V_2 + V_2^2}{2g} \\
&= \frac{(V_1 - V_2)^2}{2g} = K_L \frac{V_1^2}{2g}
\end{aligned}
\tag{3.37}
$$

Thus the loss coefficient K_L for a sudden expansion is

$$K_L = \frac{(V_1 - V_2)^2}{V_1^2} = \left(1 - \frac{V_2}{V_1}\right)^2 = \left(1 - \frac{A_1}{A_2}\right)^2 \tag{3.38}$$

3.7.2 Loss Coefficients for Conical Expansions. The head loss through a sudden expansion (Figure 3.7) can be reduced by a gradual transition, that is, by inserting a conical diffuser section.

h_f in a conical expansion $= C_L \times$ (head loss in sudden expansion)

Figure 3.7 Head loss in a conical diffuser (Streeter, V. L., *Fluid Mechanics*, © 1958, p. 188. Adapted by permission of McGraw-Hill Book Company, New York)

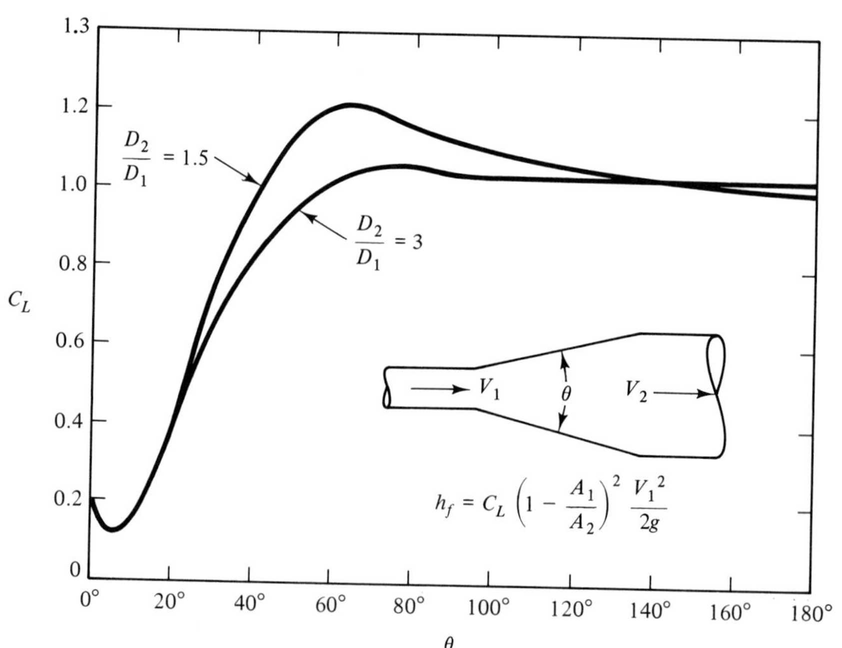

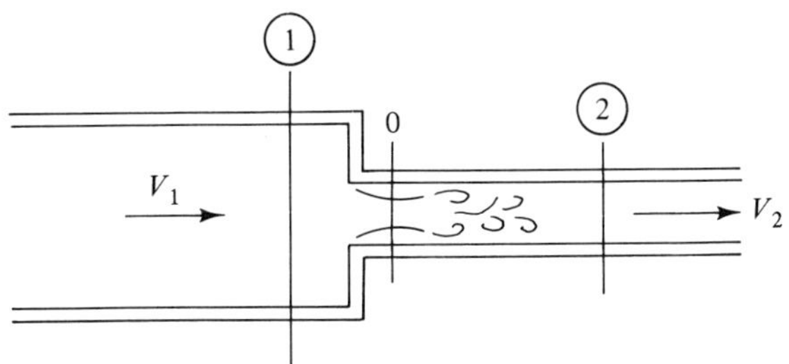

Figure 3.8 Sudden contraction

3.7.3 Head Loss in Sudden Contraction. The head loss from section 1 to the vena contracta is negligible (Figure 3.8). From the vena contracta to section 2, the loss is the same as for a sudden enlargement.

$$h_{f_{1\to2}} = h_{f_{1\to0}}^{\;0} + h_{f_{0\to2}} \tag{3.39}$$

$$= \frac{(V_0 - V_2)^2}{2g} = \left(\frac{1}{C_c} - 1\right)^2 \frac{V_2^2}{2g}$$

where $V_0 C_c A_2 = V_2 A_2$.

The contraction coefficient varies from approximately $C_c = 0.62$ to 0.68 as the area ratio varies from $A_2/A_1 = 0.1$ to 0.5.

3.7.4 Head Losses at Pipe Entrance, Exit, and Various Fittings. For minor head loss (Figure 3.9),

$$h_f = K_L \frac{V^2}{2g}$$

where V = velocity in the pipe.

Globe valve (fully open)	$K_L \approx 10$
Gate valve (fully open)	$K_L \approx 0.2$
Standard 90° elbow	$K_L \approx 0.9$
Tee (straight through)	$K_L \approx 0.4$
Tee (branch flow)	$K_L \approx 1.5$

Figure 3.9 Minor losses at exit and entrance to a reservoir

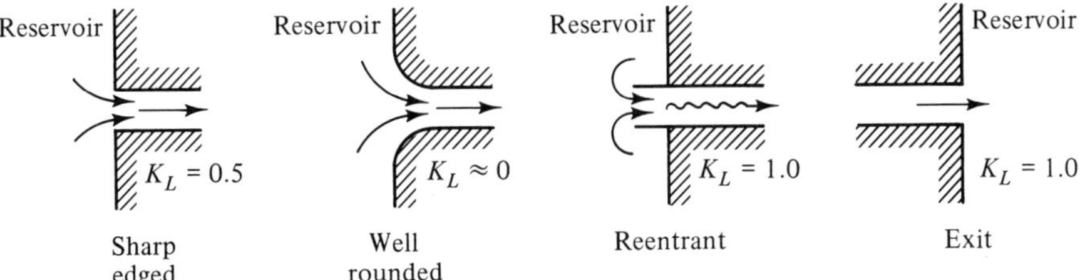

These minor head loss coefficients are relatively insensitive to Reynolds number. Furthermore, as the minor losses are generally a small part of the overall friction loss, approximate minor head loss coefficients are adequate.

3.8 Some Pipe-Flow Problems and Solutions

3.8.1. Calculate the pressure P_1 (Figure 3.10) such that 0.3 ft³/sec of water will be discharged to atmosphere. Choose points 1 and 2 as indicated in Figure 3.10 and write the energy equation as follows:

$$\frac{P_1}{\gamma} + \frac{V_1^2}{2g} + Z_1 = \frac{P_2}{\gamma} + \frac{V_2^2}{2g} + Z_2 + h_f$$

Consider each term:

$\frac{P_1}{\gamma}$ P_1 is the unknown to be determined

$\frac{V_1^2}{2g} = 0$ the tank area is large; V_1 is very small and $V_1^2 \approx 0$

$Z_1 = 20$ measured above an arbitrary datum, in this case the ground

$\frac{P_2}{\gamma} = 0$ the pressure at point 2 is equal to atmospheric pressure; we usually work with gauge pressure, so $P_2 = 0$; note that P_1, which we have to find, will also be expressed as gauge pressure

$\frac{V_2^2}{2g}$ from continuity, $Q = A_2V_2$; so

$$V_2 = \frac{Q}{A_2} = \frac{0.3}{(\pi/4)(4/12)^2} = 3.44 \text{ ft/sec}$$

$$\frac{V_2^2}{2g} = \frac{3.44^2}{64.4} = 0.184 \text{ ft}$$

$Z_2 = 10$

h_f head loss in 25 ft of 2-in. pipe + 15 ft of 4-in. pipe + minor losses, including pipe entrance, two elbows, and a sudden enlargement

Figure 3.10 Flow system with minor losses

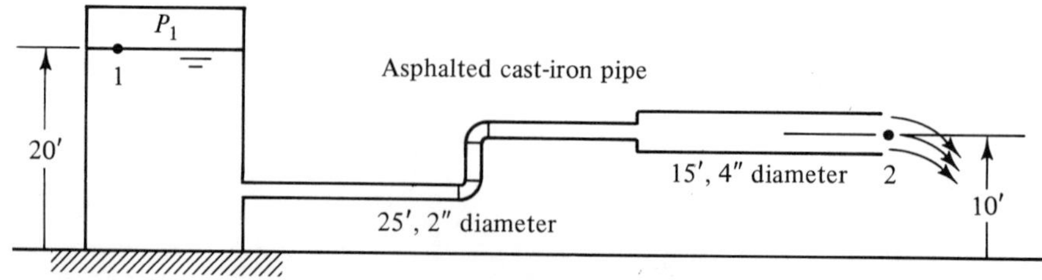

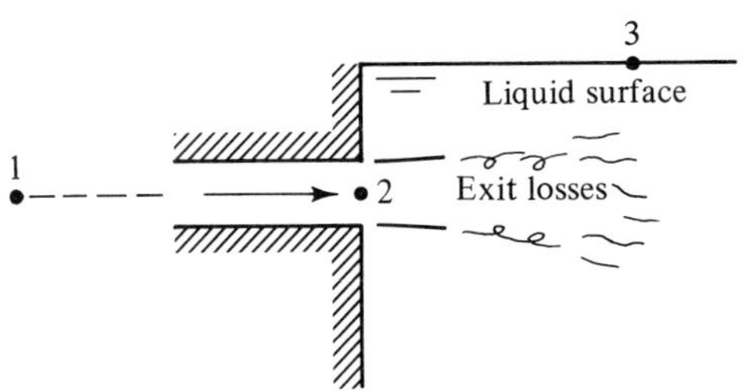

Figure 3.11 Exit loss

Note that there is no exit loss between points 1 and 2 because point 2 is located just at (before) the exit. If we had a situation as shown in Figure 3.11, then h_f from point 1 to 2 does not include exit loss, but h_f from point 1 to 3 does include exit loss.

Returning to the evaluation of head loss in our problem,

$$h_f = \left(f\frac{L}{D}\frac{V^2}{2g}\right)_{\text{2-in. pipe}} + \left(f\frac{L}{D}\frac{V^2}{2g}\right)_{\text{4-in. pipe}} + (K_1 + 2K_2 + K_3)\frac{V^2}{2g}_{\text{2-in. pipe}}$$

where $K_{1\ \text{entrance}} = 0.5$

$$K_{2\ \text{elbow}} = 0.9$$

$$K_{3\ \text{expansion}} = \left(1 - \frac{A_1}{A_2}\right)^2 = \left[1 - \left(\frac{2}{4}\right)^2\right]^2 = 0.56$$

Also, in the 2-in. diameter pipe,

$$V = \frac{Q}{A} = \frac{0.3}{(\pi/4)(2/12)^2} = 13.8 \text{ ft/sec}$$

$$\frac{V^2}{2g} = 2.94 \text{ ft}$$

$$\text{Re} = \rho\frac{VD}{\mu} = \frac{62.4 \times 13.8 \times 2/12}{2.36 \times 10^{-5} \times 32.2} = 189{,}000$$

$$\frac{\varepsilon}{D} = \frac{0.0004}{2/12} = 0.0024$$

$f = 0.0255$ from the Moody diagram.

In the 4-in.-diameter pipe,

$$V = 3.44 \text{ ft/sec}, \qquad \frac{V^2}{2g} = 0.184 \text{ ft}$$

$$\text{Re} = \frac{62.4 \times 3.44 \times 4/12}{2.36 \times 10^{-5} \times 32.2} = 94{,}200$$

$$\frac{\varepsilon}{D} = \frac{0.0004}{4/12} = 0.0012$$

$$f = 0.023$$

$$h_f = 0.0255\frac{25}{2/12} \times 2.94 + 0.023\frac{15}{4/12} \times 0.184 + (0.5 + 2 \times 0.9 + 0.56) \times 2.94$$

$$= 11.2 + 0.19 + 8.4 = 19.8\,\text{ft}$$

Returning to the energy equation,

$$\frac{P_1}{\gamma} + \frac{V_1^2}{2g} + Z_1 = \frac{P_2}{\gamma} + \frac{V_2^2}{2g} + Z_2 + h_f$$

$$\frac{P_1}{\gamma} + 0 + 20 = 0 + 0.184 + 10 + 19.8$$

$$\frac{P_1}{\gamma} = 10.0\,\text{ft}$$

$$P_1 = 62.4 \times 10 = 624\,\text{lb}_\text{f}/\text{ft}^2 = 4.3\,\text{psig}$$

3.8.2. Water at 20°C is pumped from a well to a reservoir 90 m above the well through 2 km of 30-cm-i.d. commercial steel pipe.

(a) Ignoring minor losses, find the pump power required for a flow rate of 0.2 m^3/s.
(b) If the pump is 1.8 m above the water surface in the well, and the suction pipe is 6.5 m long with one 90° elbow, calculate the water pressure at the pump inlet. (If this pressure is too near the liquid vapor pressure, cavitation will occur at the pump inlet.) Include minor losses in part b.

(a)
$$V = \frac{Q}{A} = \frac{0.2}{(\pi/4)(0.30)^2} = 2.83\ \text{m/s}$$

$$\text{Re} = \frac{\rho VD}{\mu} = \frac{998 \times 2.83 \times 0.30}{0.00101} = 839{,}000$$

For commercial steel, $\varepsilon = 0.046$ mm:

$$\frac{\varepsilon}{D} = \frac{0.0046}{30} = 0.00015$$

$$f = 0.0142 \quad \text{(Moody diagram)}$$

$$h_f = f\frac{L}{D}\frac{V^2}{2g} = 0.0142 \times \frac{2000}{0.30}\frac{(2.83)^2}{2 \times 9.81} = 38.6\ \text{m}$$

For the energy equation, choose point 1 at the liquid surface in the well and point 2 at the liquid surface in the reservoir (90 m above the well. Not shown in Figure 3.12).

$$\cancelto{0}{\frac{P_1}{\gamma}} + \cancelto{0}{\frac{V_1^2}{2g}} + Z_1 + W_{\text{pump}} = \cancelto{0}{\frac{P_2}{\gamma}} + \cancelto{0}{\frac{V_2^2}{2g}} + Z_2 + h_f$$

$$W_{\text{pump}} = Z_2 - Z_1 + h_f = 90 + 38.6 = 128.6 \text{ m}$$

$$\text{Pump power} = 128.6 \frac{\text{Nm}}{\text{N}} \times 0.2 \frac{\text{m}^3}{\text{s}} \times 998 \frac{\text{kg}}{\text{m}^3} \times 9.81 \frac{\text{N}}{\text{kg}}$$

$$= 252{,}000 \text{ Nm/s} = 252 \text{ kW}$$

(b) Energy equation between 1 and 2 in Figure 3.12:

$$\cancelto{0}{\frac{P_1}{\gamma}} + \cancelto{0}{\frac{V_1^2}{2g}} + Z_1 = \frac{P_2}{\gamma} + \frac{V_2^2}{2g} + Z_2 + h_f$$

$$\frac{P_2}{\gamma} = Z_1 - Z_2 - \frac{V_2^2}{2g} - h_f = -1.8 - \frac{V_2^2}{2g} - h_f$$

Now

$$h_f = f\frac{L}{D}\frac{V^2}{2g} + \underset{\text{Elbow}}{K_1\frac{V^2}{2g}} + \underset{\text{Entrance}}{K_2\frac{V^2}{2g}}$$

$$K_1 = 0.9 \quad \text{(elbow)}$$
$$K_2 = 1.0 \quad \text{(reentrant)}$$

and from part a

$$f = 0.0142, \qquad V = 2.83 \text{ m/s}$$

$$h_f = \left(0.0142 \times \frac{6.5}{0.030} + 0.9 + 1.0\right) \times \frac{(2.83)^2}{2 \times 9.81} = 2.03 \text{ m}$$

Figure 3.12 Water pumped from a well to a reservoir

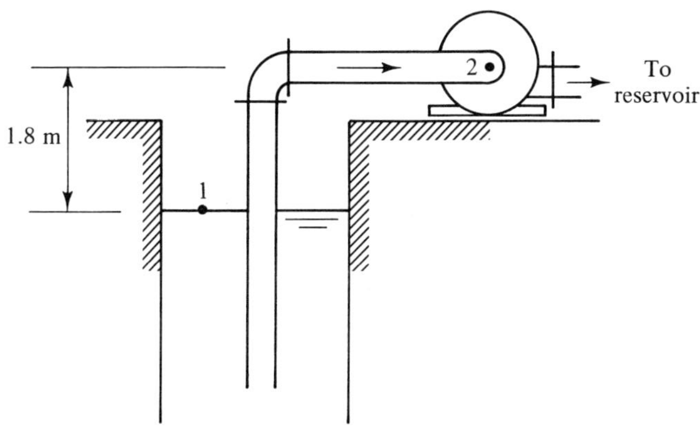

So

$$\frac{P_2}{\gamma} = -1.8 - \frac{(2.83)^2}{2 \times 9.81} - 2.03 = -4.24 \text{ m}$$

$$P_2 \text{ gauge} = -4.24\gamma = -4.24\rho g$$

$$= -4.24 \text{ m} \times 998 \text{ kg/m}^3 \times 9.81 \text{ N/kg}$$

$$= -41.5 \text{ kPa}$$

$$P_{\text{atmos}} \simeq 101 \text{ kPa}$$

$$\therefore P_2 \text{ absolute} = 101 - 41.5 = 59.5 \text{ kPa}$$

3.9 Drag and Lift

The total force exerted by a stream of fluid on a submerged body can be divided into two components (Figure 3.13). The *drag force* acts in line with the stream approaching the *submerged body*. The *lift force* acts normal to the approaching flow.

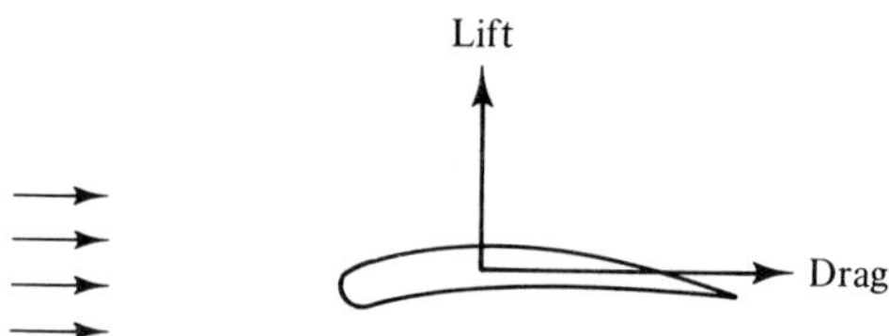

Figure 3.13 Lift and drag on an airfoil

The drag force on a two-dimensional object (long in the direction perpendicular to the flow) is the sum of two effects, pressure drag and skin friction (Figure 3.14). The total drag force is given by the expression

$$\text{Drag} = C_D \frac{\rho V^2}{2g_c} A \tag{3.40}$$

Figure 3.14 Pressure drag and skin friction drag

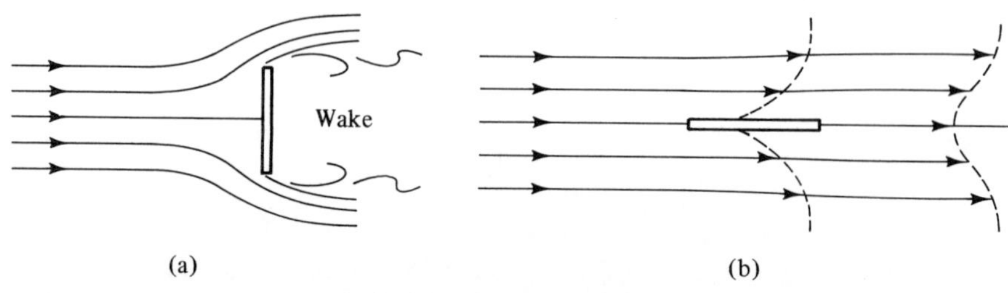

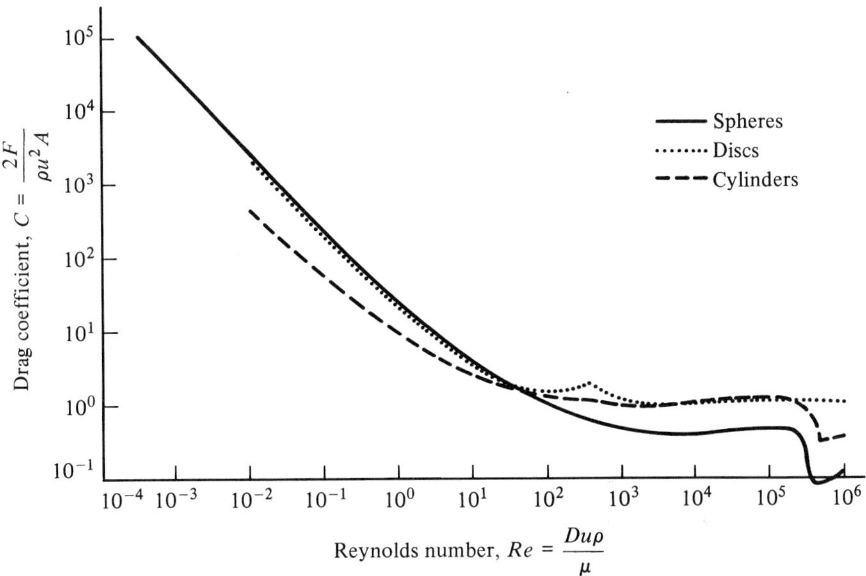

Figure 3.15 Drag coefficients for spheres, discs, and cylinders (Lapple, C. E., and C. B. Shepard, *Ind. Eng. Chem.*, *32*, 606, 1940. Reproduced by permission of the American Chemical Society)

where the drag coefficient C_D is a dimensionless constant determined experimentally for the particular shape. $\rho V^2/2g_c$ is the dynamic pressure of the approaching stream. A is usually the projected area normal to the stream, although in some cases, such as an airfoil or flat plate parallel to the stream, the area may be based on the chord length of the object. The drag coefficients for various shapes and objects are shown in Figure 3.15 and in Table 3.2. We will use these for problem solutions even if the Reynolds number is outside the range specified. This approximation is fairly good for sharp-edged objects where the flow separation point is well defined. It is less valid for rounded objects.

The lift force is due to the pressure distribution about the object, which in turn depends on the velocity distribution about the body. In Figure 3.16, the velocity over the top of the airfoil is faster than over the bottom. From the Bernoulli equation,

$$\frac{P}{\gamma} + \frac{V^2}{2g} = \text{constant}$$

we see that an increase in velocity implies a decrease in pressure. The lower pressure along the top of the body produces a lift force:

$$\text{Lift} = C_L \frac{\rho V^2}{2g_c} A \tag{3.41}$$

The area A is based on the chord length of the airfoil. The lift and drag coefficients for a typical airfoil are listed in Table 3.3.

TABLE 3.2

DRAG COEFFICIENTS FOR VARIOUS SHAPES

Two-dimensional shapes		Reynolds number	C_D	Three-dimensional shapes		Reynolds number	C_D
Circular cylinder	→ ○	10^4 to 10^5	1.2	Sphere	→ ○	10^4 to 10^5	0.47
Semitubular	→ ◖	4×10^4	1.2	Hemisphere	→ ◖	10^4 to 10^5	0.42
Semitubular	→ D	4×10^4	2.3	Hemisphere	→ D	10^4 to 10^5	1.17
Square cylinder	→ □	3.5×10^4	2.0	Cube	→ □	10^4 to 10^5	1.05
Flat plate	→ ▯	10^4 to 10^6	1.98	Cube	→ ◇	10^4 to 10^5	0.80
Elliptical cylinder	→ ⬭ 2:1	10^5	0.46	Rectangular plate with length/width = 5	→ ▯ ↕ width	10^3 to 10^5	1.20
Elliptical cylinder	→ ⬭ 8:1	2×10^5	0.20				

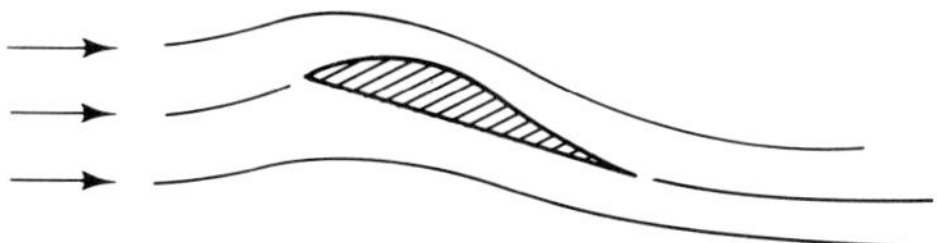

Figure 3.16 Flow over an airfoil

TABLE 3.3
LIFT AND DRAG COEFFICIENTS FOR A TYPICAL AIRFOIL

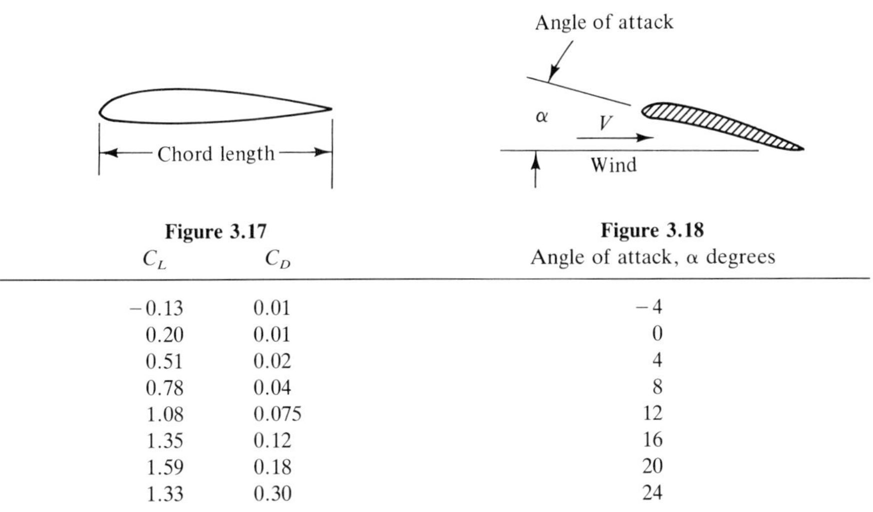

Figure 3.17 C_L	C_D	Figure 3.18 Angle of attack, α degrees
−0.13	0.01	−4
0.20	0.01	0
0.51	0.02	4
0.78	0.04	8
1.08	0.075	12
1.35	0.12	16
1.59	0.18	20
1.33	0.30	24

3.10 Preview of Two-Dimensional Fluid Flow

Our descriptive and empirical approach to lift and drag should hint at the need for a more comprehensive approach to fluid flow than the one-dimensional theory used in Chapter 2. A study of two-dimensional flow theory (in later courses) will provide more complete insight into fluid-flow behavior, though a complete analytical solution to two-dimensional problems will remain beyond the reach of undergraduates.

If it is assumed that the effects of viscosity can be ignored in flowing fluids like water or air, then the velocity and pressure distribution around submerged objects can be calculated fairly simply from potential flow theory. Examples of streamline patterns obtainable are shown in Figures 3.19 and 3.20.

In the past, potential flow theory was highly developed and capable of describing inviscid flows around many shapes. However, real flow behavior was known to be radically different than predictions from potential flow theory. For instance, the theory predicted no drag for flow around a cylinder

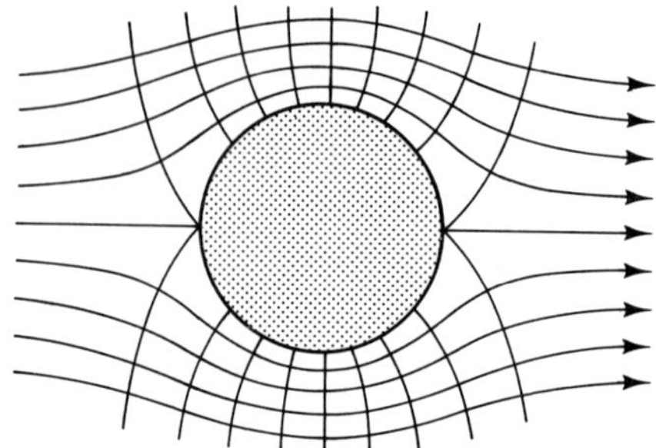

Figure 3.19 Flow around a cylinder

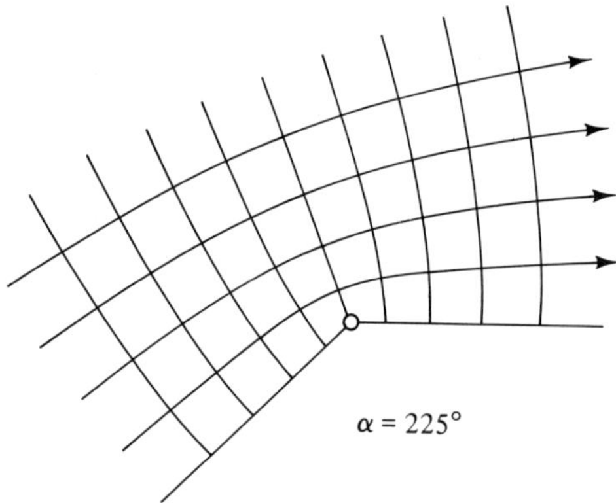

Figure 3.20 Flow around a corner

(Figure 3.19), while experience indicated the presence of a drag force and a significantly different flow pattern, particularly on the downstream side of the cylinder (Figure 3.21). Thus theoretical hydrodynamics evolved somewhat separately from the empirical art or science of hydraulics, where information on real flow behavior was assembled.

Figure 3.21 Flow across a cylinder

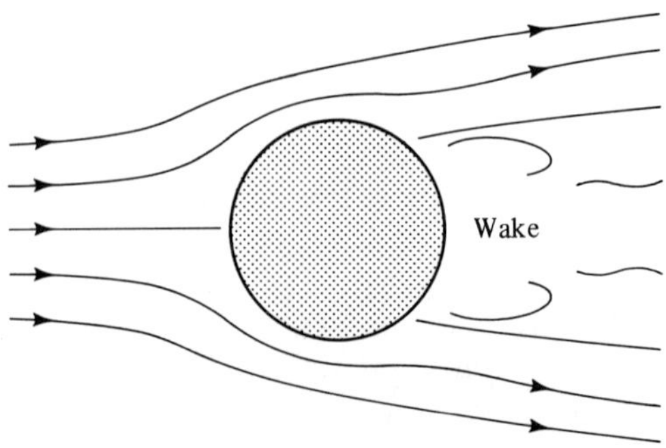

Ludwig Prandtl succeeded in merging the theoretical and the practical with his introduction of boundary layer theory. No matter how small the viscosity of a liquid, there can be no slip where the liquid is in contact with a solid boundary. Thus velocity gradients must be large in the vicinity of the solid boundary, and the viscous shear stress, $\tau = \mu(\partial V/\partial y)$, will be significant in this layer of fluid. Prandtl postulated that real flow fields could be divided into two domains for analysis: a boundary layer where the effects of viscosity must be included, and the region outside the boundary layer where potential flow theory would apply.

The presence of the viscous boundary layer has little effect in accelerating flow, where the pressure in the fluid field decreases in the flow direction. On the front half of a cylinder, for instance, the slow-moving fluid in the boundary layer just forces the steamlines to take a slightly wider path, as though the cylinder were slightly larger than it really is.

In decelerating flow, however, the boundary layer behavior has a profound effect on the overall flow pattern (as for instance just beyond the halfway point around a cylinder).

The pressure increase from A to B (Figure 3.22) is accomplished by a reduction in free-stream velocity, as shown by Bernoulli:

$$\frac{P_B - P_A}{\gamma} = \frac{V_A^2 - V_B^2}{2g} \tag{3.42}$$

However, the fluid in the boundary layer is already moving more slowly and does not have sufficient kinetic energy to negotiate the rise in pressure. This causes a back flow in the boundary layer near the wall, and the main stream separates from the wall. There is no further pressure recovery after separation, and thus the pressure on the back side in the region of the wake does not balance the pressure on the front side of the cylinder. This causes the pressure drag on the cylinder.

Figure 3.22 Boundary layer separation

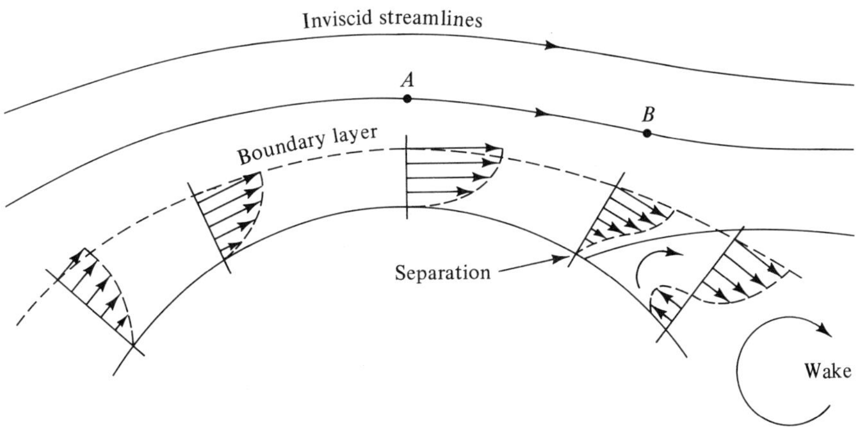

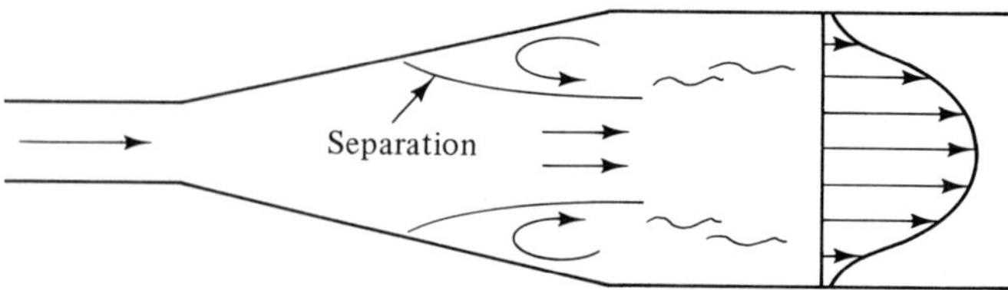

Figure 3.23 Diffuser flow with separation

The same reasoning also applies to head losses through nozzles and diffusers. In the nozzle (accelerating flow), the boundary layer is well behaved and the losses are small. In the diffuser (decelerating flow, increasing pressure), the boundary layer may cause separation and mixing with large losses (Figure 3.23).

PROBLEMS

3.1. A 75-cm-diameter pipe, 1 km long, is fed with water from a reservoir whose surface is 15 m above the pipe exit. The pipe is of cast iron, is clean, and makes a rounded connection to the reservoir. Estimate the rate of discharge from the pipe. *Answer:* 1.62 m^3/s

3.2. Two water reservoirs differing in elevation by 30 m are connected by 3 km of 30-cm cast-iron pipe. What is the flow in m^3/s between the two reservoirs? *Answer:* 0.124 m^3/s

3.3. In a certain plant a chemical solution used in the absorption tower is pumped continuously from the bottom of the storage tank through a 4-in. steel pipe (4.026 in. i.d.) and then through the spray head at the top of the tower. The cross-sectional area of the spray head is large compared to the pipe. The depth of solution in the storage tank is 5 ft, and the vertical distance from the bottom of this tank to the spray head is 105 ft. In a test on this equipment the following data were recorded: flow rate = 200 U.S. gal/min; SG of solution = 1.10; friction loss from tank to spray head = 15 ft; spray head pressure = 5 psig; power input to the pump = 14.0 hp (1 U.S. gal = 0.1337 ft^3; 1 hp = 33,000 ft lb/min). What is the efficiency of the pump? *Answer:* 50%

3.4. A steel pipe with an inside diameter of 340 mm is to be designed to carry 8000 m^3 of oil/24-hr day from an oil field to a refinery located 300 km from the source. The difference in elevation between the two ends of the pipe is negligible.
(a) Estimate the power required to overcome friction in the pipe line.
(b) The maximum allowable pressure in any section of the line is 4.5 MPa. Additional pumping stations will therefore have to be inserted in the line at suitable intervals. What is the smallest number of pumping stations required: *Data:* At the average temperature involved, the oil has an absolute viscosity of 50 cp and a specific gravity of 0.87 (100 cp = 1 poise = 1 g/cm-s). *Answer:* (a) 1.29 MW; (b) four stations

3.5. A pump at an elevation of 300 m is pumping water at the rate of 2.7 m^3/min through 2000 m of 15-cm-i.d. steel pipe to a reservoir at an elevation of 420 m. What pressure will be found in the pipe 800 m from the pump where the elevation is 340 m? *Answer:* 1.22 MPa gauge

3.6. From a reservoir whose surface is at an elevation of 750 ft, water is pumped through 4000 ft of 12-in. steel pipe across a valley to a second reservoir whose surface is at an elevation of 800 ft. If, during pumping, the pressure is 80 psig at a point in the pipe midway of its length and at an elevation of 650 ft, estimate the rate of discharge and the power required by the pump. *Answer:* 6.88 ft^3/sec, 93 hp

3.7. If 6800 liters of water/min is to be discharged from a cast-iron pipe, 305 m long, which has its discharge end 1.20 m below the surface of the supplying reservoir, what is the required pipe diameter? *Answer:* 0.353 m

3.8. Water is to be pumped through 300 m of piping to a filtration plant located 25 m above the river (source). The pressure rise across the pump is not to exceed 350 kPa. The line contains one open gate valve and two 90° bends. If the flow rate is to be 20 m^3/min, find the required diameter for the asphalt-dipped cast-iron pipe. *Answer:* 0.351 m

3.9. A 48-in. main, carrying 75.4 ft^3 of water/sec branches at a point A into two pipes, one 2000 ft long and 3 ft in diameter, and the other 6000 ft long and 2 ft in diameter. Both pipes come together at a point B and continue as a single 48-inch pipe. Determine the flow in each of the branch pipes (assume the pipes are cast iron). *Answer:* 63.0 and 12.4 ft^3/sec

3.10. Using dimensional analysis with four basic dimensions (F, M, L, t), it was shown in Section 3.4 that the friction factor f is a function of Reynolds number and the roughness ratio ε/D. Perform your own analysis using a dimensional system of three basic dimensions (M, L, t) to show that the friction factor is a function of Reynolds number and roughness ratio.

3.11. Two open cylindrical tanks are connected by 300 m of 80-mm-i.d. iron pipe laid horizontally. Reservoir A is 8 m in diameter and its water level is 12 m above that in reservoir B, whose diameter is 5 m. How long, after opening a valve on the pipeline, will it be before the reservoir levels are the same: Assume f to remain constant, and neglect minor losses. *Answer:* 12.5 hr

3.12. Oil (SG 0.934, viscosity 1.65 poise) flows through 6000 m of 15 cm steel pipe with an average velocity of 1.2 m/s. Calculate the lost head due to friction. *Answer:* 184 m

3.13. Oil (density 50 lb/ft^3, viscosity 0.80 poise) flows from a reservoir through 50 ft of $\frac{1}{4}$-in. drawn tubing. What is the flow in ft^3/sec and the Reynolds number if the outlet from the tube is 40 ft below the oil surface in the reservoir? *Answer:* 0.000111 ft^3/sec; Re = 6.29

3.14. What is the pressure drop in the line if 740 Imperial gallons of gasoline/min flow through a horizontal pipe 200 mm in diameter and 13 km long? Specific gravity and kinematic viscosity of gasoline are 0.75 and 0.01 cm^2/sec, respectively. Relative roughness of the pipe is 0.0004. (1 Imp gal = 1.2 U.S. gal = 4.54 liters.) *Answer:* 1.36 MPa

3.15. Carbon dioxide flows through a horizontal commercial steel pipe 10 cm in diameter. At inlet the absolute pressure is 925 kPa, the temperature 40°C, and the average velocity 12 m/s. What is the pressure drop in 45 m of pipe? Assume the flow to be incompressible. $\mu = 15.9 \times 10^{-6}$ kg/m-s. *Answer:* 8.6 kPa

3.16. The oil from a cutting machine collects in a pan and drains away to a reservoir through a vertical $\frac{3}{8}$-in. diameter tube. What is the maximum flow of oil that can be handled? Specific gravity and viscosity of the oil are 0.85 and 4×10^{-4} ft²/sec, respectively. *Answer:* 0.0019 ft³/sec

3.17. A viscous fluid flows down a long sloping tray under steady-state conditions. The tray is wide so that edge effects can be neglected (Figure 3.24).

(a) Considering a strip of fluid w units wide (into the paper), derive an expression for the velocity gradient dV/dy at height y. Note that the fluid pressures at sections AA and BB are equal, so the shear force acting on the plane at y is balanced by the weight component of the shaded fluid acting down the plane.

(b) Derive an expression for the fluid velocity at height y by integrating your expression for dV/dy.

(c) Derive an expression for the volume flow rate of fluid down the tray per unit width of tray (into the paper). *Answer:* $\frac{1}{3}(\gamma/\mu)\, h^3 \sin\theta$ vol. units per unit width per sec.

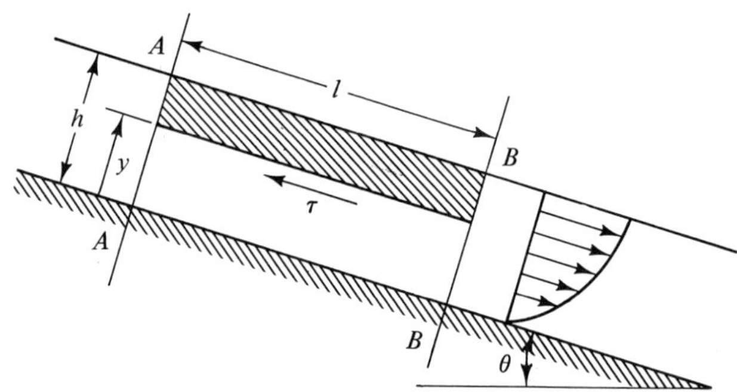

Figure 3.24 Flow down a flat plate

3.18. A 15-cm pipe carrying water suddenly enlarges to a diameter of 30 cm. The velocity of flow in the larger pipe is 2 m/s. Calculate the lost head in meters of water and the power loss in kilowatts. What is the change in static pressure from the small pipe exit to a downstream location where the velocity has become uniform? (Neglect the pipe wall friction.) *Answer:* $h_f = 1.84$ m; 2.55 kW; 12 kPa increase

3.19. A level pipeline carrying water is abruptly enlarged from 4 to 8 in. in diameter. The velocity in the 4-in. pipe is 16 ft/sec. How much power is dissipated at the enlargement? If the pressure head in the 4-in. pipe is 50 ft, what must it be in the 8-in. pipe after the velocity has become uniform? Ignore any wall friction in this region. *Answer:* 0.35 hp; 51.5 ft

3.20. Repeat Problem 3.18 with the abrupt enlargement replaced by a conical diffuser with an included angle of $\theta = 10°$. How long is the conical transition section from the 15-cm pipe to the 30-cm pipe? *Answer:* h_f = 0.275; 0.383 kW; 27 kPa, length = 86 cm

3.21. Water flows due to gravity only through 1.5 km of 80-mm-i.d. steel pipe from a mountain lake to a water reservoir on a farm below. The lake surface is 100 m above the farm reservoir.

(a) Including the effects of pipe friction, calculate the maximum flow rate of water to the reservoir.

(b) The farmer's son urges his father to install a water turbine at the pipe exit to generate electric power for the farm. What is the maximum power that could be generated, and what would be the water flow rate with the turbine installed? *Answer:* (a) 0.0118 m^3/s; (b) 4.33 kW, 0.0066 m^3/s

3.22. A rectangular flat plate 1.2 m × 0.25 m is immersed in a stream of water running with a velocity of 5 km/hr. Calculate the approximate force developed against the plate when held normal to the flow. *Answer:* 347 N

3.23. A spherical storage tank 10 m in diameter is to be monitored in the open several meters above the ground. It is required that the supports be designed so that the tank will not be moved in a 150 km/hr wind. What is the expected total force on the tank? Take the air temperature as 20°C and the pressure as 1 atm. *Answer:* 38.5 kN

3.24. What would be the approximate force caused by a gale of 70 mph against a vertical billboard 10 ft high by 40 ft long? The temperature and pressure of the air are 70°F and 1 atm, respectively. *Answer:* 5900 lb_f

3.25. A semitubular cylinder of radius 0.25 m and length 10 m is submerged in a stream of water moving at 3 m/s. The concave side of the cylinder faces upstream.

(a) Calculate the drag force acting on the cylinder.

(b) What is the drag force if the convex side of the cylinder faces upstream? *Answer:* (a) 51.6 kN; (b) 26.9 kN

3.26. **(a)** A streamlined car with frontal area of 18 ft^2 travels at 70 mph. What horsepower is required to overcome the air drag?

(b) If the car travels 70 mph against a 20-mph headwind, what horsepower increase is required due to the headwind? Assume C_D = 0.3. *Answer:* (a) 12.6 hp; (b) 8.2 hp increase

Fluid Flow in Boundary Layers 4

4.1 Introduction to Boundary Layer Flow

In 1904, the German scientist Ludwig Prandtl suggested that fluid flow could be divided into two regions, a thin film or boundary layer adjacent to a solid surface where fluid viscosity is important, and the remainder of the flow field

Figure 4.1 Fluid flow over a flat plate

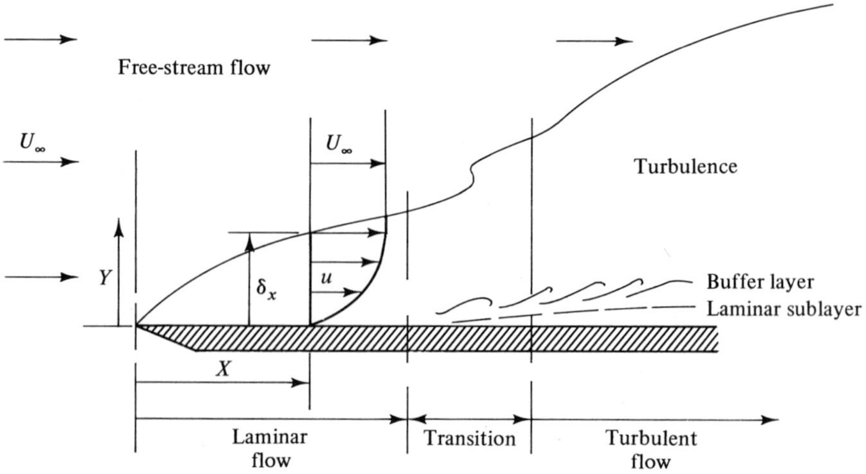

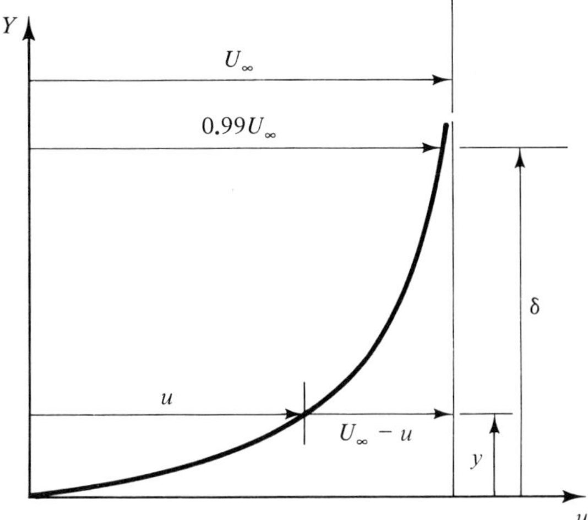

Figure 4.2 Boundary layer velocity profile at some location x

where friction effects can be neglected. Some familiarity with boundary layer behavior will extend our understanding of fluid flow and will be essential later for an understanding of convective heat and mass transfer.

Consider a fluid with a free stream velocity U_∞ flowing parallel to a flat plate. The boundary layer development is illustrated in Figure 4.1. Initially, the layer of fluid near the plate surface flows in laminar fashion. As the boundary layer develops, the laminar flow becomes unstable and changes to turbulent flow. Even in this turbulent region, the flow immediately adjacent to the solid surface remains laminar.

At any location x along the plate, the boundary layer is described by its thickness δ, the shape of the velocity profile (as the velocity reduces from U_∞ in the main stream to zero at the plate surface), and whether the boundary layer flow is laminar or turbulent (Figure 4.2). Because the boundary layer velocity merges gradually into the main stream, the thickness δ is not well defined. The boundary layer thickness δ is usually chosen as the distance from the solid surface to the point where the velocity is $0.99U_\infty$. Other so-called thicknesses indicate the reductions in mass flow, momentum, and kinetic energy caused by the boundary layer.

4.2 Displacement Thickness δ_1 (Mass-Flow Deficit)

The mass-flow deficit caused by velocity reduction in the boundary layer is $\int_0^\infty \rho(U_\infty - u)dy$. Consider this amount of mass flowing at free-stream velocity through a passage of depth δ_1.

$$\text{Mass-flow deficit} = \rho_\infty U_\infty \delta_1 = \int_0^\infty \rho(U_\infty - u)\, dy$$

Assuming the flow to be incompressible ($\rho = \rho_\infty$), then

$$\delta_1 = \int_0^\infty \frac{\rho(U_\infty - u)\, dy}{\rho_\infty U_\infty} = \int_0^\infty (1 - \frac{u}{U_\infty})\, dy \tag{4.1}$$

4.3 Momentum Thickness δ_2 (Momentum Deficit)

The momentum deficit caused by velocity reduction in the boundary layer is $\int_0^\infty \rho u(U_\infty - u)\, dy$. Consider the equivalent momentum flowing at free-stream velocity through a passage of depth δ_2.

$$\text{Momentum deficit} = (\rho_\infty U_\infty \delta_2) U_\infty = \int_0^\infty \rho u(U_\infty - u)\, dy$$

For incompressible flow,

$$\delta_2 = \frac{\int_0^\infty \rho u(U_\infty - u)\, dy}{\rho_\infty U_\infty^2} = \int_0^\infty \frac{u}{U_\infty}\left(1 - \frac{u}{U_\infty}\right) dy \tag{4.2}$$

4.4 Kinetic Energy Thickness δ_3 (Kinetic Energy Deficit)

The kinetic energy deficit caused by velocity reduction in the boundary layer is

$$\int_0^\infty \rho u\left(\frac{U_\infty^2 - u^2}{2}\right) dy$$

Consider the equivalent kinetic energy flowing at free-stream velocity through a passage of depth δ_3.

$$\text{Kinetic-energy deficit} = (\rho_\infty U_\infty \delta_3)\frac{U_\infty^2}{2} = \int_0^\infty \rho u\left(\frac{U_\infty^2 - u^2}{2}\right) dy$$

For incompressible flow,

$$\delta_3 = \frac{\int_0^\infty \rho u(U_\infty^2 - u^2)\, dy}{\rho_\infty U_\infty \times U_\infty^2} = \int_0^\infty \frac{u}{U_\infty}\left[1 - \left(\frac{u}{U_\infty}\right)^2\right] dy \qquad (4.3)$$

4.5 Notes Regarding the Boundary Layer Thickness and Integration Limits

As illustrated in Figure 4.2, the boundary layer velocity profile merges gradually with the mainstream velocity U_∞. Thus the definition for the boundary layer thickness δ must be arbitrary (i.e., where $u \simeq 0.99U_\infty$). To account for all the "missing" velocity, it is appropriate to use infinity as the upper limit on y when integrating over the real boundary layer profile.

However, when we introduce some function $u = f(y)$ as a realistic approximation for the boundary layer velocity profile, this function $u = f(y)$ becomes equal to U_∞ at some precise value of y, which defines the boundary layer thickness δ. The function $u = f(y)$ approximates the velocity profile in the range from $y = 0$ to $y = \delta$, but not outside this range. When performing integrations using $u = f(y)$, we must use δ rather than infinity as the upper limit on y.

It is often convenient to use a new variable $\eta = y/\delta$ to define location in the boundary layer nondimensionally.

4.6 Summary of Boundary Layer Thickness for Incompressible Flow

Boundary layer thickness:

δ = height where $u = 0.99U_\infty$ in real velocity profile

= height where $u = U_\infty$ when using the approximation $u = f(y)$ for the velocity profile

Displacement thickness:

$$\delta_1 = \int_{y=0}^{\infty}\left(1 - \frac{u}{U_\infty}\right) dy$$

$$\simeq \int_{y=0}^{\delta}\left(1 - \frac{u}{U_\infty}\right) dy \simeq \delta \int_{\eta=0}^{1}\left(1 - \frac{u}{U_\infty}\right) d\eta \qquad (4.1)$$

Momentum thickness:

$$\delta_2 = \int_{y=0}^{\infty} \frac{u}{U_\infty}\left(1 - \frac{u}{U_\infty}\right) dy$$
$$\simeq \int_{y=0}^{\delta} \frac{u}{U_\infty}\left(1 - \frac{u}{U_\infty}\right) dy \simeq \delta \int_{\eta=0}^{1} \frac{u}{U_\infty}\left(1 - \frac{u}{U_\infty}\right) d\eta \tag{4.2}$$

Kinetic energy thickness:

$$\delta_3 = \int_{y=0}^{\infty} \frac{u}{U_\infty}\left[1 - \left(\frac{u}{U_\infty}\right)^2\right] dy$$
$$\simeq \int_{y=0}^{\delta} \frac{u}{U_\infty}\left[1 - \left(\frac{u}{U_\infty}\right)^2\right] dy \simeq \delta \int_{\eta=0}^{1} \frac{u}{U_\infty}\left[1 - \left(\frac{u}{U_\infty}\right)^2\right] d\eta \tag{4.3}$$

Example 4.6.1

Find the displacement thickness, momentum thickness, and energy thickness as fractions of the boundary layer thickness for the velocity profile $u = U_\infty \sin(\pi/2)(y/\delta)$ and find the shear stress at the wall (Figure 4.3).

Displacement thickness:

$$\delta_1 = \delta \int_{y=0}^{y=\delta} \left(1 - \frac{u}{U_\infty}\right) d\left(\frac{y}{\delta}\right)$$
$$= \delta \int_0^1 \left(1 - \sin\frac{\pi}{2}\eta\right) d\eta$$
$$= \delta\left(\eta + \frac{2}{\pi}\cos\frac{\pi}{2}\eta\right)_0^1 = \frac{\pi - 2}{\pi}\delta$$
$$\frac{\delta_1}{\delta} = \frac{\pi - 2}{\pi} = 0.3634$$

Figure 4.3 Sine curve representing velocity profile ($0 \leq y \leq \delta$)

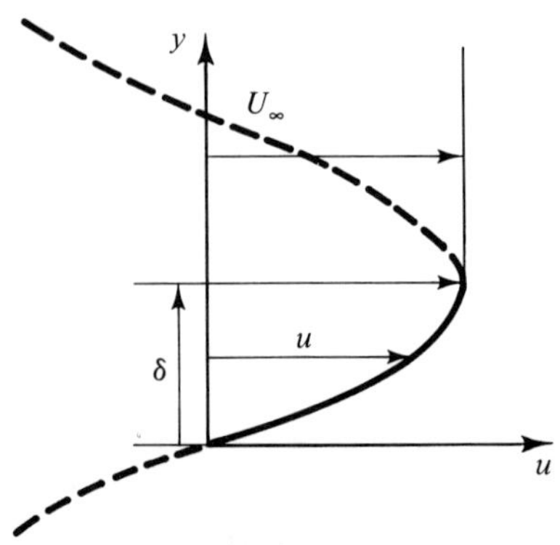

Momentum thickness:

$$\frac{\delta_2}{\delta} = \int_0^1 \frac{u}{U_\infty}\left(1 - \frac{u}{U_\infty}\right) d\eta$$

$$= \int_0^1 \sin\frac{\pi}{2}\eta\, d\eta - \int_0^1 \sin^2\frac{\pi}{2}\eta\, d\eta$$

$$= -\frac{2}{\pi}\cos\frac{\pi}{2}\eta\,\Big|_0^1 - \int_0^1 \left(\frac{1}{2} - \frac{\cos\pi\eta}{2}\right) d\eta$$

$$= \frac{2}{\pi} - \frac{1}{2} = \frac{4-\pi}{2\pi} = 0.1366$$

Energy thickness:

$$\frac{\delta_3}{\delta} = \int_0^1 \frac{u}{U_\infty}\left[1 - \left(\frac{u}{U_\infty}\right)^2\right] d\eta$$

$$= \int_0^1 \sin\frac{\pi}{2}\eta\, d\eta - \int_0^1 \sin^3\frac{\pi}{2}\eta\, d\eta$$

$$= \frac{2}{\pi} + \frac{2}{\pi}\int_{\eta=0}^1 \left(1 - \cos^2\frac{\pi}{2}\eta\right) d\left(\cos\frac{\pi}{2}\eta\right)$$

$$= \frac{2}{\pi} + \frac{2}{\pi}\left[\cos\frac{\pi}{2}\eta\,\Big|_0^1 - \frac{\cos^3\frac{\pi}{2}\eta}{3}\,\Big|_0^1\right]$$

$$= \frac{2}{\pi}\left(1 - \frac{2}{3}\right) = \frac{2}{3\pi} = 0.2122$$

Shear stress at the wall:

$$\tau_w = \mu\frac{\partial u}{\partial y}\Big|_{\text{at } y=0}$$

$$= \frac{\mu U_\infty}{\delta}\,\frac{\partial\left(\frac{u}{U_\infty}\right)}{\partial(y/\delta)}\Bigg|_{\text{at } y/\delta=0}$$

$$= \frac{\mu U_\infty}{\delta}\,\frac{\partial\left[\sin\frac{\pi}{2}\eta\right]}{\partial\eta}\Bigg|_{\text{at } \eta=0}$$

$$= \frac{\mu U_\infty}{\delta}\times\frac{\pi}{2}\cos\frac{\pi}{2}\eta\Big|_{\text{at } \eta=0}$$

$$= \frac{\pi\mu U_\infty}{2\delta}$$

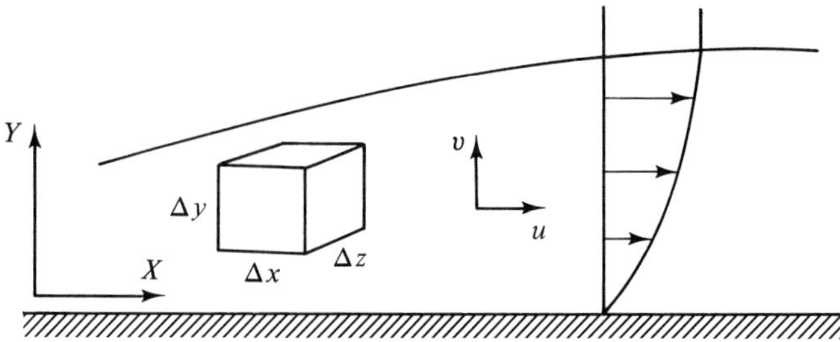

Figure 4.4 Elemental control volume $\Delta x\Delta y\Delta z$ within the boundary layer

4.7 Boundary Layer Equations

To evaluate the changes taking place in the boundary layer in the flow direction is our next objective. For a specific physical setup operating under particular conditions, it is conceivable that the velocity profiles could be measured at successive locations (values of x) along the solid surface. However, an analytical approach offers the prospect of acquiring information that is more general and comprehensive in character. Thus we seek to develop mathematical statements describing the fluid motion in the boundary layer.

In Chapter 2, where we dealt with one-dimensional fluid flow, we presented the concepts of mass, momentum, and energy conservation (see Section 2.7). It seems logical that we should once again explore the concepts of mass, momentum, and energy as applied to steady flow in boundary layers.

Consider a control volume $\Delta x\Delta y\Delta z$ fixed in space within the boundary layer (Figure 4.4). Fluid flows through and around this control volume with velocity u in the x direction and velocity v in the y direction.

4.8 Mass Conservation Equation

For steady flow, with no accumulation of mass within the control volume, the requirement for mass conservation can be stated as follows:

Rate of mass leaving − rate of mass entering = 0

or, restated,

Excess of mass leaving compared to mass entering = 0

Mathematically, referring to Figure 4.5,

$$\frac{\partial}{\partial x}(\rho u)\,\Delta x\Delta y\Delta z + \frac{\partial}{\partial y}(\rho v)\,\Delta y\Delta x\Delta z = 0 \tag{4.4}$$

Now cancel out the volume $\Delta x\Delta y\Delta z$,

$$\frac{\partial}{\partial x}(\rho u) + \frac{\partial}{\partial y}(\rho v) = 0 \tag{4.5}$$

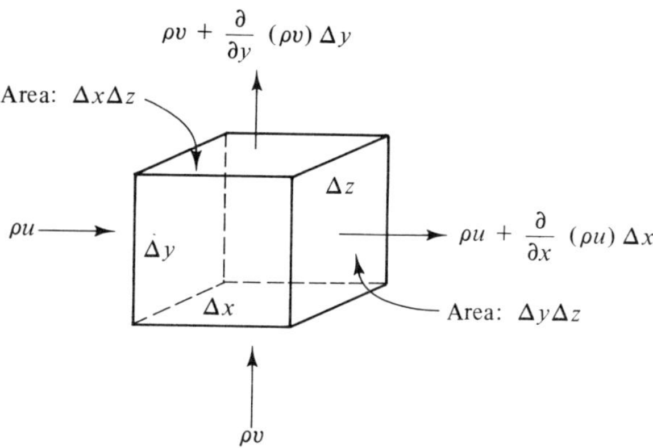

Figure 4.5 Mass fluxes through the control volume

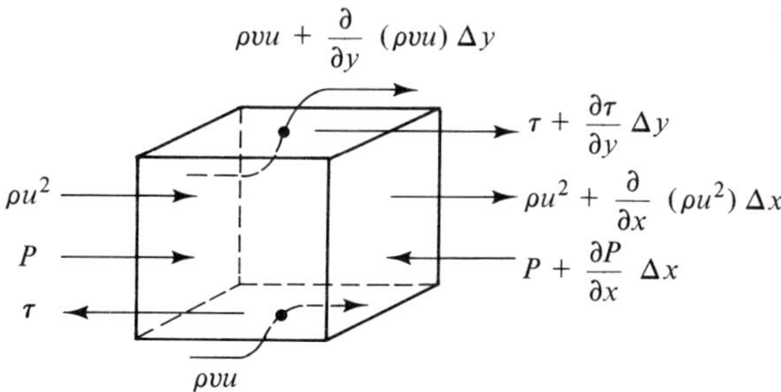

Figure 4.6 Momentum fluxes and surface stresses

For incompressible flow, ρ is constant:

$$\frac{\partial u}{\partial x} + \frac{\partial v}{\partial y} = 0 \tag{4.6}$$

This is the mass conservation equation for incompressible two-dimensional flow.

4.9 Momentum Equation

Referring to Figure 4.6,

$$\text{Linear momentum in } x \text{ direction} = \text{mass} \times \text{velocity} = mu$$

$$\text{Rate of flow of momentum} = \dot{m}u$$

$$\text{Momentum flux} = \frac{\text{rate of flow of momentum}}{\text{area}} = \frac{\dot{m}u}{A}$$

$$x\text{-direction momentum flux through left face} = \frac{(\rho A u)u}{A} = \rho u^2$$

$$x\text{-direction momentum flux through bottom face} = \frac{(\rho A v)u}{A} = \rho v u$$

In steady flow, the net rightward force acting on the volume element equals the rate of increase of rightward momentum:

$$-\frac{\partial P}{\partial x}\Delta x \Delta y \Delta z + \frac{\partial \tau}{\partial y}\Delta y \Delta x \Delta z = \frac{\partial}{\partial x}(\rho u^2)\,\Delta x \Delta y \Delta z + \frac{\partial}{\partial y}(\rho v u)\,\Delta y \Delta x \Delta z \tag{4.7}$$

Thus, for incompressible flow,

$$-\frac{\partial P}{\partial x} + \frac{\partial \tau}{\partial y} = \rho\left(\frac{\partial u^2}{\partial x} + \frac{\partial v u}{\partial y}\right)$$

$$-\frac{\partial P}{\partial x} + \frac{\partial}{\partial y}\left(\mu \frac{\partial u}{\partial y}\right) = \rho\left(2u\frac{\partial u}{\partial x} + v\frac{\partial u}{\partial y} + u\frac{\partial v}{\partial y}\right)$$

From the mass conservation equation, substitute $\partial v/\partial y = -\partial u/\partial x$ and in addition treat μ as a constant.

The boundary layer momentum equation becomes

$$u\frac{\partial u}{\partial x} + v\frac{\partial u}{\partial y} = -\frac{1}{\rho}\frac{\partial P}{\partial x} + \frac{\mu}{\rho}\frac{\partial^2 u}{\partial y^2} \tag{4.8}$$

4.10 Discussion of Boundary Conditions

With the derivation of the momentum equation fresh in our minds, it is instructive to look at the boundary conditions.

At $y = 0$, the concept of viscosity requires that there be no slip at a solid surface (i.e., $u = 0$ at $y = 0$). Also, no mass flows into or out of the solid surface (i.e., $v = 0$ at $y = 0$). Thus the momentum equation simplifies to

$$\frac{\partial P}{\partial x} = \mu\frac{\partial^2 u}{\partial y^2}$$

at the solid surface. For flow over a flat plate $\partial P/\partial x = 0$, and thus $\partial^2 u/\partial y^2 = 0$ at the flat plate surface. We will use this boundary condition in a later section.

Outside the boundary layer (at $y = \infty$) the velocity u equals the free-stream velocity U_∞. There are no further changes in u for changing values of y (i.e., $\partial u/\partial y = 0$ and $\partial^2 u/\partial y^2 = 0$). The momentum equation now becomes

$$U_\infty\frac{\partial U_\infty}{\partial x} = -\frac{1}{\rho}\frac{\partial P}{\partial x}$$

or

$$\frac{1}{2}\frac{\partial U_\infty^2}{\partial x} = -\frac{1}{\rho}\frac{\partial P}{\partial x}$$

or

$$\frac{dP}{\rho} + \frac{dU_\infty^2}{2} = 0$$

This is simply Euler's equation for frictionless flow at constant elevation.

4.11 Energy Equation

At the top face of the element (Figure 4.7), a shear force acting in the direction of motion does work *on* the volume element. At the bottom face, a shear force acting opposite to the direction of flow *extracts* work from the element. For steady state in the volume element

$$\begin{pmatrix}\text{Net work input}\\ \text{per unit time}\\ \text{by friction}\end{pmatrix} = \begin{pmatrix}\text{net rate of energy}\\ \text{convected}\\ \text{out}\end{pmatrix} + \begin{pmatrix}\text{net rate of}\\ \text{heat transfer}\\ \text{out}\end{pmatrix}$$

$$\frac{\partial}{\partial y}(\tau u)\,\Delta y\Delta x\Delta z = \frac{\partial}{\partial x}(\dot{m}_x h_T)\,\Delta x + \frac{\partial}{\partial y}(\dot{m}_y h_T)\,\Delta y + \frac{\partial}{\partial y}(q_y)\,\Delta y \tag{4.9}$$

where h_T represents total head enthalpy, which is

$$h_T = h + ke = h + \frac{u^2 + v^2}{2} \simeq h + \frac{u^2}{2},$$

since $v^2 << u^2$.

Note: h = specific enthalpy of the fluid. Also,

$$\dot{m}_x = \rho u\,\Delta y\Delta z$$

$$\dot{m}_y = \rho v\,\Delta x\Delta z$$

$$q_y = -k\,\Delta x\Delta z\,\frac{\partial T}{\partial y}$$

Inserting these expressions for $\dot{m}_x$, $\dot{m}_y$, and q_y and canceling the product $\Delta x\Delta y\Delta z$, we have for the energy equation

$$\frac{\partial}{\partial y}(\tau u) = \frac{\partial}{\partial x}(\rho u h_T) + \frac{\partial}{\partial y}(\rho v h_T) + \frac{\partial}{\partial y}\left(-k\,\frac{\partial T}{\partial y}\right) \tag{4.10}$$

Note that

$$\begin{aligned}\frac{\partial}{\partial x}(\rho u h_T) + \frac{\partial}{\partial y}(\rho v h_T) &= h_T\left[\frac{\partial}{\partial x}(\rho u) + \frac{\partial}{\partial y}(\rho v)\right] + \rho u\,\frac{\partial h_T}{\partial x} + \rho v\,\frac{\partial h_T}{\partial y}\\ &= \rho u\,\frac{\partial h}{\partial x} + \rho v\,\frac{\partial h}{\partial y} + \rho u\,\frac{\partial(u^2/2)}{\partial x} + \rho v\,\frac{\partial(u^2/2)}{\partial y}\end{aligned}$$

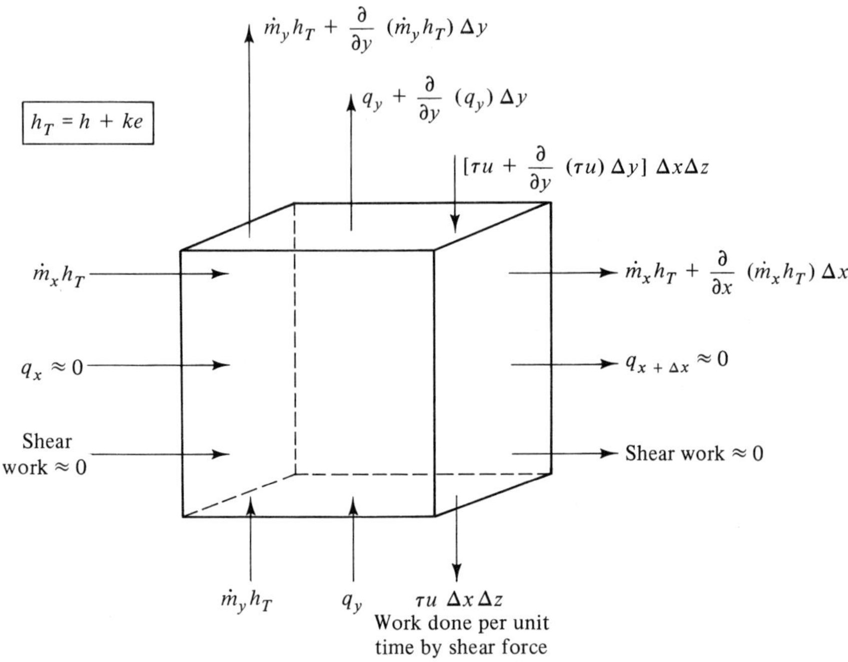

Figure 4.7 Energy flows

From equation (4.5),

$$h_T\left[\frac{\partial}{\partial x}(\rho u) + \frac{\partial}{\partial y}(\rho v)\right] = 0$$

and assuming constant thermal conductivity k,

$$\frac{\partial}{\partial y}\left(-k\frac{\partial T}{\partial y}\right) = -k\frac{\partial^2 T}{\partial y^2}$$

At this stage in the energy equation development, we have

$$\frac{\partial}{\partial y}(\tau u) = \rho u\frac{\partial h}{\partial x} + \rho v\frac{\partial h}{\partial y} + \rho u\frac{\partial(u^2/2)}{\partial x} + \rho v\frac{\partial(u^2/2)}{\partial y} - k\frac{\partial^2 T}{\partial y^2} \tag{4.11}$$

Part of the net work input to the volume element by shear stresses is dissipated within by friction, and part may cause a change in kinetic energy. Expand the left side of the energy equation so that

$$\frac{\partial}{\partial y}(\tau u) = \tau\frac{\partial u}{\partial y} + u\frac{\partial \tau}{\partial y} = \mu\left(\frac{\partial u}{\partial y}\right)^2 + u\mu\frac{\partial^2 u}{\partial y^2} \tag{4.12}$$

Now recollect the momentum equation (4.8):

$$u\frac{\partial u}{\partial x} + v\frac{\partial u}{\partial y} = -\frac{1}{\rho}\frac{\partial P}{\partial x} + \frac{\mu}{\rho}\frac{\partial^2 u}{\partial y^2} \tag{4.8}$$

Multiply the momentum equation by ρu, while noting that $u\,\partial u = \partial(u^2/2)$:

$$\rho u\,\frac{\partial(u^2/2)}{\partial x} + \rho v\,\frac{\partial(u^2/2)}{\partial y} = -u\,\frac{\partial P}{\partial x} + u\mu\,\frac{\partial^2 u}{\partial y^2} \tag{4.13}$$

Replace the term $u\mu(\partial^2 u/\partial y^2)$ in the equation for shear work (4.12) with its derived value from the momentum equation (4.13).

$$\frac{\partial}{\partial y}(\tau u) = \mu\left(\frac{\partial u}{\partial y}\right)^2 + \rho u\,\frac{\partial(u^2/2)}{\partial x} + \rho v\,\frac{\partial(u^2/2)}{\partial y} + u\,\frac{\partial P}{\partial x} \tag{4.14}$$

Substitute the value given for $(\partial/\partial y)(\tau u)$ in equation (4.14) into equation (4.11) and cancel the kinetic energy terms.

$$\underbrace{\rho u\,\frac{\partial h}{\partial x} + \rho v\,\frac{\partial h}{\partial y}}_{\text{Enthalpy change}} = \underbrace{k\,\frac{\partial^2 T}{\partial y^2}}_{\text{Heat transfer}} + \underbrace{\mu\left(\frac{\partial u}{\partial y}\right)^2}_{\text{Frictional dissipation}} + \underbrace{u\frac{\partial P}{\partial x}}_{\text{Pressure effects}} \tag{4.15}$$

It comes as no surprise that the enthalpy of the flowing fluid is affected by heat transfer and frictional dissipation. It may be less obvious why the pressure gradient should affect the enthalpy, particularly since pressure gradient does not appear in the thermodynamic first law statement for a control volume.

$$\delta q = dh + dke + dPe + \delta W_{\text{out}}$$

or rearranged for comparison with equation (4.15), assuming no change in potential energy,

$$\underbrace{dh}_{\text{Enthalpy change}} = \underbrace{\delta q}_{\text{Heat transfer}} - \underbrace{d\text{ke}}_{\text{Kinetic energy}} + \underbrace{\delta W_{\text{in}}}_{\text{Work input}} \tag{4.16}$$

Comparing the boundary layer energy equation (4.15) with the first law statement (4.16), we see that the pressure and dissipation terms are equivalent to kinetic energy and work effects. Also, recollecting that the term $(\partial/\partial y)(\tau u)$ in equation (4.14) is work input, this equation states quite explicitly the connection between work input, kinetic energy dissipation, and pressure effects.

Substituting for $\partial h = C_P\,\partial T$, the boundary layer energy equation can be expressed as

$$\rho u C_P\,\frac{\partial T}{\partial x} + \rho v\,C_P\,\frac{\partial T}{\partial y} = k\,\frac{\partial^2 T}{\partial y^2} + \mu\left(\frac{\partial u}{\partial y}\right)^2 + u\,\frac{\partial P}{\partial x} \tag{4.17}$$

Except for high-speed flow under strong pressure gradients, we can ignore the term $u(\partial P/\partial x)$ and write the energy equation for either compressible or incompressible flow as

$$u\,\frac{\partial T}{\partial x} + v\,\frac{\partial T}{\partial y} = \frac{k}{\rho C_P}\,\frac{\partial^2 T}{\partial y^2} + \frac{\mu}{\rho C_p}\left(\frac{\partial u}{\partial y}\right)^2 \tag{4.18}$$

4.12 Summary of Boundary Layer Equations for Incompressible Flow

Continuity $$\frac{\partial u}{\partial x} + \frac{\partial v}{\partial y} = 0 \tag{4.6}$$

Momentum $$u\,\frac{\partial u}{\partial x} + v\,\frac{\partial u}{\partial y} = -\frac{1}{\rho}\frac{\partial P}{\partial x} + \nu\,\frac{\partial^2 u}{\partial y^2} \tag{4.8}$$

Energy $$u\,\frac{\partial T}{\partial x} + v\,\frac{\partial T}{\partial y} = \frac{k}{\rho C_p}\frac{\partial^2 T}{\partial y^2} + \frac{\mu}{\rho C_P}\left(\frac{\partial u}{\partial y}\right)^2 \tag{4.18}$$

Blasius solved these equations exactly for laminar boundary layer flow over a flat plate. Some other exact solutions exist, giving the boundary layer profile, the boundary layer thickness, the frictional drag, and so on. However, given some arbitrary flow situation, it is extremely unlikely that an exact solution exists or is obtainable. Thus approximate methods have been developed for getting rough solutions to real problems. The degree of approximation involved in the approximate methods can be evaluated by comparison with a few available exact solutions.

4.13 Von Kármán Momentum Integral

When the partial differential equations for continuity and momentum are solved, the motion of every fluid element in the boundary layer is determined. Because of the mathematical difficulties associated with exact solutions, approximate methods are sought, capable of adequate accuracy with a minimum of complication. Instead of accounting for the motion of every particle, the momentum integral method deals approximately with the whole boundary layer thickness as it develops in the flow direction.

Consider the forces and momentum changes in the x direction for steady flow associated with the control volume 1234 shown in Figure 4.8. Note that

Figure 4.8 Momentum integral method

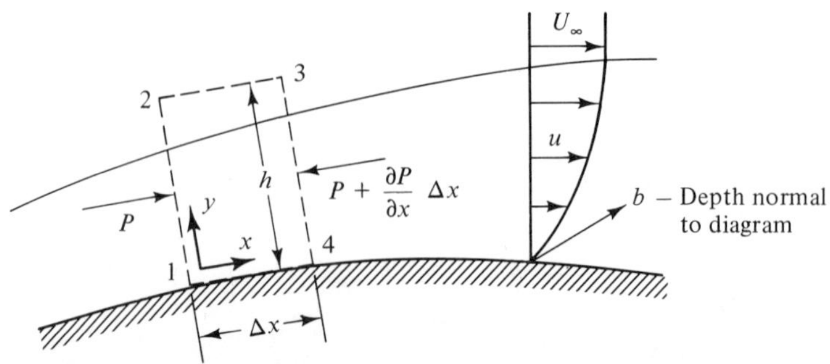

pressure is constant in the y direction but may change in the x direction.

$$\text{Mass flow rate into left side (12)} = b\int_0^h \rho u\, dy$$

Mass flow rate out of right side (34)

$$= b\int_0^h \rho u\, dy + b\frac{\partial}{\partial x}\left(\int_0^h \rho u\, dy\right)\Delta x$$

Conservation of mass requires that,

$$\text{Mass flow rate out the top (23)} = -b\frac{\partial}{\partial x}\left(\int_0^h \rho u\, dy\right)\Delta x$$

$$\text{Rightward momentum flow rate into left side (12)} = b\int_0^h \rho u^2\, dy$$

Rightward momentum flow rate out right side (34)

$$= b\int_0^h \rho u^2\, dy + b\frac{\partial}{\partial x}\left(\int_0^h \rho u^2\, dy\right)\Delta x$$

Rightward momentum flow rate out the top (23)

$$= (\dot{m}_{\text{out top}}) \times U_\infty = -U_\infty b\frac{\partial}{\partial x}\left(\int_0^h \rho u\, dy\right)\Delta x$$

Net increase of rightward momentum flow rate

$$= b\frac{\partial}{\partial x}\left(\int_0^h \rho u^2\, dy\right)\Delta x - U_\infty b\frac{\partial}{\partial x}\left(\int_0^h \rho u\, dy\right)\Delta x$$

$$\text{Net rightward pressure force} = -bh\left(\frac{\partial P}{\partial x}\right)\Delta x$$

$$\text{Rightward shear force at base (14)} = -\tau_w b\,\Delta x$$

$$\text{Net rightward force} = -\tau_w b\,\Delta x - bh\frac{\partial P}{\partial x}\Delta x$$

Now set the rightward force equal to the increase in rightward momentum flow rate and cancel out the product $b\,\Delta x$:

$$\frac{\partial}{\partial x}\left(\int_0^h \rho u^2\, dy\right) - U_\infty\frac{\partial}{\partial x}\left(\int_0^h \rho u\, dy\right) = -\tau_w - h\frac{\partial P}{\partial x} \tag{4.19}$$

Note that in the (inviscid) free-stream region

$$\frac{dP}{\rho} + U_\infty\, dU_\infty = 0$$

so that

$$h\frac{\partial P}{\partial x} = -\rho h U_\infty \frac{\partial U_\infty}{\partial x}$$

Now h is equal to $\int_0^h dy$, ρ is a constant, and while U_∞ varies with x it is not a function of y. Thus $\rho h U_\infty$ can be written $\int_0^h \rho U_\infty\, dy$ so that

$$h\frac{\partial P}{\partial x} = -\left(\int_0^h \rho U_\infty\, dy\right)\frac{\partial U_\infty}{\partial x}$$

Substitute this expression for $h(\partial P/\partial x)$ into equation (4.19):

$$\frac{\partial}{\partial x}\left(\int_0^h \rho u^2\, dy\right) - U_\infty\frac{\partial}{\partial x}\left(\int_0^h \rho u\, dy\right) = -\tau_w + \left(\int_0^h \rho U_\infty\, dy\right)\frac{\partial U_\infty}{\partial x} \tag{4.20}$$

Using the calculus relation $A\, dB = d(AB) - B\, dA$, express the term

$$U_\infty\frac{\partial}{\partial x}\left(\int_0^h \rho u\, dy\right) = \frac{\partial}{\partial x}\left(U_\infty\int_0^h \rho u\, dy\right) - \left(\int_0^h \rho u\, dy\right)\frac{\partial U_\infty}{\partial x}$$

When we move the last term in equation (4.20) to the left side, we get

$$-\frac{\partial}{\partial x}\left(U_\infty\int_0^h \rho u\, dy - \int_0^h \rho u^2\, dy\right) - \left(\int_0^h \rho U_\infty\, dy - \int_0^h \rho u\, dy\right)\frac{\partial U_\infty}{\partial x} = -\tau_w \tag{4.21}$$

For incompressible flow,

$$\frac{\partial}{\partial x}\left[U_\infty^2\int_0^h \frac{u}{U_\infty}\left(1-\frac{u}{U_\infty}\right)dy\right] + \left[U_\infty\int_0^h\left(1-\frac{u}{U_\infty}\right)dy\right]\frac{\partial U_\infty}{\partial x} = \frac{\tau_w}{\rho} \tag{4.22}$$

Since the quantity $1 - (u/U_\infty)$ is zero in the region from $y = \delta$ outward, the upper limit of integration can be changed from h to δ (or ∞) without changing the value of the integrals. These integrals should be recognized as the momentum thickness δ_2 and the displacement thickness δ_1. Thus

$$\frac{\partial}{\partial x}(U_\infty^2\delta_2) + U_\infty\delta_1\frac{\partial U_\infty}{\partial x} = \frac{\tau_w}{\rho}$$

The quantities involved vary with x but not with y, so we can replace the partial derivatives with total derivatives.

$$\frac{d}{dx}(U_\infty^2\delta_2) + U_\infty\delta_1\frac{dU_\infty}{dx} = \frac{\tau_w}{\rho} \tag{4.23}$$

This is the von Kármán momentum integral equation for boundary layer flow with a pressure gradient. It can be used for both laminar and turbulent boundary layers.

For flow over a flat plate with no pressure gradient, the momentum equation becomes easy to use. In this case, U_∞ is constant, $dU_\infty/dx = 0$, and the momentum integral equation reduces to

$$\frac{d\delta_2}{dx} = \frac{\tau_w}{\rho U_\infty^2} \tag{4.24}$$

Furthermore, it is known that such a boundary layer has similar velocity profiles, growing thicker as we move in the flow direction. To solve for the boundary layer growth, we assume a reasonable velocity profile shape in the form $u/U_\infty = f(y/\delta) = f(\eta)$ and use equation (4.24) to calculate the rate of growth of δ.

4.14 Laminar Boundary Layer Growth on a Flat Plate

To illustrate the use of the momentum integral equation for a flat plate, assume that the velocity profile can be represented as part of a sine curve (see Figure 4.3):

$$\frac{u}{U_\infty} = \sin\frac{\pi}{2}\frac{y}{\delta} = \sin\frac{\pi}{2}\eta \tag{4.25}$$

Note that for this function

$$\text{At } y = 0: \qquad \frac{u}{U_\infty} = 0, \qquad \frac{\partial^2(u/U_\infty)}{\partial\eta^2} = 0$$

$$\text{At } y = \delta: \qquad \frac{u}{U_\infty} = 1, \qquad \frac{\partial(u/U_\infty)}{\partial\eta} = 0$$

We have shown (Section 4.7) that, for the velocity profile $u/U_\infty = \sin(\pi/2)\eta$,

$$\delta_2 = \frac{4-\pi}{2\pi}\delta \quad \text{and} \quad \tau_w = \frac{\pi\mu U_\infty}{2\delta}$$

We may now solve the momentum integral equation:

$$\frac{d\delta_2}{dx} = \frac{\tau_w}{\rho U_\infty^2}$$

$$\frac{d}{dx}\left(\frac{4-\pi}{2\pi}\delta\right) = \frac{\pi\mu U_\infty}{2\delta} \times \frac{1}{\rho U_\infty^2} = \frac{\pi\mu}{2\delta\rho U_\infty} \tag{4.24}$$

$$\delta\frac{d\delta}{dx} = \frac{\pi^2}{4-\pi}\frac{\mu}{\rho U_\infty}$$

$$\frac{\delta^2}{2} = \frac{\pi^2}{4-\pi}\frac{\mu x}{\rho U_\infty} + C$$

Setting $\delta = 0$ at $x = 0$, the integration constant C is zero and the boundary layer thickness is

$$\delta = \sqrt{\frac{2\pi^2}{4 - \pi}} \sqrt{\frac{\mu x}{\rho U_\infty}} = 4.795 \sqrt{\frac{\mu x}{\rho U_\infty}} \tag{4.26}$$

Another common way of expressing this result is

$$\frac{\delta}{x} = 4.795 \sqrt{\frac{\mu}{\rho U_\infty x}} = \frac{4.795}{\sqrt{\mathrm{Re}_x}} \tag{4.27}$$

Having assumed a velocity profile shape and after solving for the boundary layer thickness, we can now solve for the viscous drag. The shear stress at any location x is given by

$$\begin{aligned} \tau_w &= \frac{\pi}{2} \frac{\mu U_\infty}{\delta} \\ &= \frac{\pi}{2} \frac{\mu U_\infty}{4.795} \sqrt{\frac{\rho U_\infty}{\mu x}} \\ &= 0.3276 \sqrt{\frac{\mu \rho U_\infty{}^3}{x}} \end{aligned} \tag{4.28}$$

Consider a rectangular surface of width b and length l:

$$\begin{aligned} \text{Total viscous drag} &= \int \tau_w dA \\ &= b \int_{x=0}^{x=l} \tau_w dx \\ &= 0.3276b \int_0^l \sqrt{\frac{\mu \rho U_\infty{}^3}{x}}\, dx \\ &= 0.655b \sqrt{\mu \rho U_\infty{}^3 l} \end{aligned} \tag{4.29}$$

Another approach to the calculation of viscous drag utilizes the concept of momentum thickness. The momentum deficit caused by velocity reduction in the boundary layer is $\rho U_\infty^2 \times b\delta_2$, which is also equal to the drag force.

$$\begin{aligned} \text{Total drag} &= \rho U_\infty{}^2 b \delta_2 \\ &= \rho U_\infty{}^2 b \left(\frac{4 - \pi}{2\pi} \times \delta \right) \\ &= \rho U_\infty{}^2 b \frac{4 - \pi}{2\pi} \times 4.795 \sqrt{\frac{\mu l}{\rho U_\infty}} \\ &= 0.655b \sqrt{\mu \rho U_\infty{}^3 l} \end{aligned} \tag{4.30}$$

The drag coefficient is defined by the relation

$$\text{Drag} = C_D A \frac{\rho U_\infty^2}{2}$$

$$C_D = \frac{\text{Drag}}{A} \frac{2}{\rho U_\infty^2}$$

The local drag coefficient (at location x on the plate) is

$$C_{Dx} = \tau_{wx} \frac{2}{\rho U_\infty^2} = 0.655 \sqrt{\frac{\mu}{\rho U_\infty x}} = \frac{0.655}{\sqrt{\text{Re}_x}} \tag{4.31}$$

The average coefficient of drag for the whole plate is

$$\begin{aligned} C_{D\text{Plate}} &= \frac{\text{total drag}}{\text{plate area}} \times \frac{2}{\rho U_\infty^2} \\ &= \frac{0.655 b \sqrt{\mu \rho U_\infty^3 l}}{bl} \times \frac{2}{\rho U_\infty^2} \\ &= 1.31 \sqrt{\frac{\mu}{\rho U_\infty l}} = \frac{1.31}{\sqrt{\text{Re}_l}} \end{aligned} \tag{4.32}$$

4.15 Choice of a Suitable Velocity Profile

To select a suitable function to describe the boundary layer velocity profile, we look at the boundary conditions that should be satisfied by that function. At $y = 0$, we have $u = 0$, and for flow over a flat plate with $\partial P/\partial x = 0$, it follows that $\partial^2 u/\partial y^2 = 0$ (see Section 4.11).

At $y = \delta$, we have $u = U_\infty$ and $\partial u/\partial y = 0$ and the higher derivatives $\partial^n u/\partial y^n = 0$.

In general, when a function is chosen to represent the velocity profile, it does not satisfy all the boundary conditions. Even if all the boundary conditions were satisfied, the function chosen might not be "correct." It might not match the real profile in the region between $y = 0$ and $y = \delta$. This is what makes the von Kármán integral method an approximate method.

We have just performed the calculations for boundary layer thickness and drag using a sine wave for the velocity function. Another approach is to choose a polynomial with as many constants as boundary conditions to be satisfied.

$$\frac{u}{U_\infty} = \underbrace{\underbrace{\underbrace{a + b\eta}_{\text{1st degree}} + c\eta^2}_{\text{2nd degree}} + d\eta^3}_{\text{3rd degree polynomial}} + \cdots \tag{4.33}$$

4.15.1 First-degree Polynomial. If we want a function that gives $u/U_\infty = 0$ at $\eta = 0$ and $u/U_\infty = 1$ at $\eta = 1$, without necessarily satisfying any of the other boundary conditions, then a first-degree polynomial will suffice.

$$\frac{u}{U_\infty} = a + b\eta$$

$$\text{At } \eta = 0, \quad \frac{u}{U_\infty} = 0; \qquad \text{therefore } a = 0$$

$$\text{At } \eta = 1, \quad \frac{u}{U_\infty} = 1; \qquad \text{therefore } b = 1$$

Thus

$$\frac{u}{U_\infty} = \eta \tag{4.34}$$

4.15.2 Second-degree polynomial

$$\frac{u}{U_\infty} = a + b\eta + c\eta^2$$

$$\text{At } \eta = 0, \quad \frac{u}{U_\infty} = 0; \qquad \text{therefore} \qquad a = 0$$

$$\text{At } \eta = 1, \quad \frac{u}{U_\infty} = 1; \qquad \text{therefore} \qquad b + c = 1$$

$$\text{At } \eta = 1, \quad \frac{\partial(u/U_\infty)}{\partial\eta} = 0; \qquad \text{therefore } \underline{b + 2c = 0}$$

$$c = -1, b = +2$$

Thus

$$\frac{u}{U_\infty} = 2\eta - \eta^2 \tag{4.35}$$

4.15.3 Third-degree Polynomial

$$\frac{u}{U_\infty} = a + b\eta + c\eta^2 + d\eta^3$$

$$\frac{\partial(u/U_\infty)}{\partial\eta} = b + 2c\eta + 3d\eta^2$$

$$\frac{\partial^2(u/U_\infty)}{\partial\eta^2} = 2c + 6d\eta$$

$$\text{At } \eta = 0, \quad \frac{u}{U_\infty} = 0; \quad \text{therefore} \quad a = 0$$

$$\text{At } \eta = 0, \quad \frac{\partial^2(u/U_\infty)}{\partial \eta^2} = 0; \quad \text{therefore } c = 0$$

$$\text{At } \eta = 1, \quad \frac{u}{U_\infty} = 1; \quad \text{therefore} \quad b + d = 1$$

$$\text{At } \eta = 1, \quad \frac{\partial(u/U_\infty)}{\partial \eta} = 0; \quad \text{therefore} \quad \underline{b + 3d = 0}$$

$$d = -\tfrac{1}{2}, \; b = +\tfrac{3}{2}$$

Thus

$$\frac{u}{U_\infty} = \frac{3}{2}\eta - \frac{1}{2}\eta^3 \tag{4.36}$$

TABLE 4.1
LAMINAR BOUNDARY LAYER PROPERTIES ON A FLAT PLATE BASED ON VARIOUS ASSUMED PROFILE SHAPES

Velocity profile function	$\delta_1\sqrt{\dfrac{U_\infty}{\nu x}}$	$\delta_2\sqrt{\dfrac{U_\infty}{\nu x}}$	$C_{Dx}\sqrt{\dfrac{U_\infty x}{\nu}}$
Blasius exact	1.72	0.664	0.664
$\frac{u}{U_\infty} = \sin\frac{\pi}{2}\eta$	1.74	0.654	0.654
$\frac{u}{U_\infty} = \eta$	1.73	0.577	0.577
$\frac{u}{U_\infty} = 2\eta - \eta^2$	1.83	0.730	0.730
$\frac{u}{U_\infty} = \frac{3}{2}\eta - \frac{1}{2}\eta^3$	1.74	0.646	0.646

Table 4.1 shows that even the straight-line assumption for the velocity profile ($u/U_\infty = \eta$) produces reasonably accurate results for the laminar boundary layer properties.

4.16 Turbulent Boundary Layer Shape and Wall Shear Stress

To apply the von Kármán momentum integral, we need a function to describe the turbulent boundary layer profile and a way to evaluate the wall shear stress. J. Nikuradse amassed and presented a wealth of experimental data on flow through pipes, resulting in the following equation for the turbulent velocity profile in pipes:

$$\frac{u}{U_{\mathrm{CL}}} = \left(\frac{y}{R}\right)^{1/n} \tag{4.37}$$

where u = velocity at location y

U_{CL} = mean turbulent velocity at pipe centerline

y is measured from pipe wall toward centerline

R = internal pipe radius

n = exponent, which varies slightly with Re

For most purposes we can ignore the slight variation of the exponent and use the $\frac{1}{7}$ power law to describe the profile.

$$\frac{u}{U_{\mathrm{CL}}} = \left(\frac{y}{R}\right)^{1/7}$$

Furthermore, it has been found that this same power law describes the turbulent velocity profile for external boundary layer flow if we substitute U_∞ for U_{CL} and δ for R. So the first requirement for application of the von Kármán integral equation to turbulent boundary layer flow is satisfied; we have a suitable function to describe the velocity profile.

$$\frac{u}{U_\infty} = \left(\frac{y}{\delta}\right)^{1/7} = \eta^{1/7} \tag{4.38}$$

Since we understand that even in turbulent boundary layers there is a laminar sublayer, we might be tempted to calculate the shear stress at the wall using Newton's law of friction ($\tau_w = \mu(\partial u/\partial y)|_{y=0}$) and the $\frac{1}{7}$ power law to evaluate $(\partial u/\partial y)|_{y=0}$. When we proceed along this line, we find that

$$\left.\frac{\partial u}{\partial y}\right|_{y=0} = \text{infinity}$$

It is obvious that the shear stress at the wall cannot be infinite, so we must conclude that the $\frac{1}{7}$ power law, which describes the overall profile well, cannot be valid in the immediate vicinity of the wall. We must use some other method to get the wall shear stress, so we turn once again to the experimental data on turbulent flow in pipes to learn about the shear stress at the wall. Consider first a cylindrical plug of fluid of radius r and length l inside a smooth pipe of radius R. In steady flow, the shear stress along the plug sides must balance the pressure force driving the plug of fluid.

$$\pi r^2(P_1 - P_2) = \tau\, 2\pi r l$$

Thus the shear stress for either laminar or turbulent flow in a pipe can be expressed as

$$\tau = \frac{P_1 - P_2}{l} \frac{r}{2}$$

It might be noted here that the intensity of the shearing stress in a fluid flowing through a pipe is proportional to the distance r measured from the centerline. At the pipe wall, we have

$$\tau_w = \frac{P_1 - P_2}{L} \frac{R}{2} \tag{4.39}$$

where L = pipe length. Thus the wall shear stress can be determined from the pressure gradient. Now the pressure loss in a pipe is given by equation (3.21):

$$h_f = f \frac{L}{D} \frac{\overline{V}^2}{2g} \tag{3.21}$$

where h_f = head loss = $(P_1 - P_2)/\gamma$

f = friction factor

$\overline{V}$ = average velocity (Q/A)

D = pipe diameter

Equation (3.21) rearranged is

$$\frac{P_1 - P_2}{L} = f \times \frac{\gamma}{2R} \times \frac{\overline{V}^2}{2g} = \frac{f\rho\overline{V}^2}{4R} \tag{4.40}$$

Substitute equation (4.40) into the shear stress equation (4.39):

$$\tau_w = \tfrac{1}{8} f\rho\overline{V}^2 \tag{4.41}$$

In 1911, Blasius reviewed the experimental data for turbulent flow through smooth pipes and established the following relation:

$$f = 0.3164 \left(\frac{\overline{V}D}{\nu}\right)^{-1/4} = \frac{0.3164}{\mathrm{Re}^{0.25}} \tag{4.42}$$

The Blasius relation is valid for Reynolds numbers up to 100,000. With the aid of the Blasius relation for f, we can now express the wall shear stress in terms of the flow parameters.

$$\tau_w = \frac{1}{8}\left[0.3164 \left(\frac{\overline{V}D}{\nu}\right)^{-1/4}\right]\rho\overline{V}^2 \tag{4.43}$$

We want the shear to be expressed in terms of U_∞ instead of $\overline{V}$. Recollect that U_∞ for boundary layer flow is analogous to U_{CL} for pipe flow. For the $\frac{1}{7}$ power law profile, the average velocity $\overline{V}$ is equal to 0.817 U_{CL}. Making the substitution in the shear equation,

$$\tau_w = \frac{1}{8}\left[0.3164\left(\frac{0.817U_\infty \times 2R}{\nu}\right)^{-1/4}\right]\rho(0.817U_\infty)^2$$

Then substituting δ for R, the resulting expression for wall shear stress in turbulent boundary layer flow is

$$\tau_w = 0.023\rho U_\infty^{\,2}\left(\frac{U_\infty\delta}{\nu}\right)^{-1/4} = \frac{0.023\rho U_\infty^{\,2}}{\mathrm{Re}_\delta^{\,0.25}} \qquad (4.44)$$

4.17 Turbulent Boundary Layer Growth on a Flat Plate

It is our intention now to calculate the rate of growth of the turbulent boundary layer and the drag associated with turbulent boundary layer flow. We begin with the von Kármán momentum integral (4.24) for flow with no pressure gradient.

$$\frac{d\delta_2}{dx} = \frac{\tau_w}{\rho U_\infty^{\,2}} \qquad (4.24)$$

The equation for shear stress at the wall (4.44) is

$$\frac{\tau_w}{\rho U_\infty^{\,2}} = 0.023\left(\frac{U_\infty\delta}{\nu}\right)^{-1/4} \qquad (4.44)$$

Substituting the $\frac{1}{7}$ power law profile into equation (4.2) defining the momentum thickness, we have

$$\delta_2 = \delta\int_{\eta=0}^{1}\frac{u}{U_\infty}\left(1-\frac{u}{U_\infty}\right)d\eta = \delta\int_0^1 \eta^{1/7}(1-\eta^{1/7})\,d\eta = \frac{7}{72}\delta \qquad (4.45)$$

Using these expressions for wall shear and momentum thickness in the von Kármán equation (4.24), we obtain

$$\frac{d}{dx}\left(\frac{7}{72}\delta\right) = 0.023\left(\frac{U_\infty\delta}{\nu}\right)^{-1/4}$$
$$\delta^{1/4}\frac{d\delta}{dx} = \frac{72}{7}\times 0.023\left(\frac{\nu}{U_\infty}\right)^{1/4} \qquad (4.46)$$

Integrating,

$$\delta^{5/4} = \frac{5}{4}\times\frac{72}{7}\times 0.023\left(\frac{\nu}{U_\infty}\right)^{1/4}x + C$$
$$= 0.296\left(\frac{\nu}{U_\infty}\right)^{1/4}x + C \qquad (4.47)$$

If we assume that the turbulent boundary layer begins at the leading edge of the plate (i.e., $\delta = 0$ at $x = 0$), then the integration constant is zero, and we have for the turbulent boundary layer thickness

$$\delta = 0.38 \left(\frac{\nu}{U_\infty}\right)^{1/5} x^{4/5} \tag{4.48}$$

Expressed in another form, we have

$$\frac{\delta}{x} = 0.38 \left(\frac{\nu}{U_\infty x}\right)^{1/5} = \frac{0.38}{(\mathrm{Re}_x)^{1/5}} \tag{4.49}$$

To get the drag related to turbulent flow, we begin with the shear stress equation (4.44):

$$\begin{aligned} \tau_w &= 0.023\rho U_\infty^2 \left(\frac{U_\infty \delta}{\nu}\right)^{-1/4} \\ &= 0.023\rho U_\infty^2 \left\{\frac{U_\infty}{\nu}\left[0.38\left(\frac{\nu}{U_\infty}\right)^{1/5} x^{4/5}\right]\right\}^{-1/4} \\ &= 0.029\rho U_\infty^2 \left(\frac{\nu}{U_\infty x}\right)^{1/5} \end{aligned} \tag{4.50}$$

Considering a rectangular plate of width b and length l,

$$\begin{aligned} \text{Drag} &= \int \tau_w \, dA \\ &= b \times 0.029\rho U_\infty^2 \left(\frac{\nu}{U_\infty}\right)^{1/5} \int_{x=0}^{x=l} x^{-1/5}\, dx \\ &= 0.036 b\rho U_\infty^2 \left(\frac{\nu}{U_\infty}\right)^{1/5} l^{4/5} \\ &= 0.036\rho U_\infty^2 bl \left(\frac{\nu}{U_\infty l}\right)^{1/5} \end{aligned} \tag{4.51}$$

The local drag coefficient is given by

$$C_{D_x} = \tau_w \frac{2}{\rho U_\infty^2} = \frac{0.058}{(\mathrm{Re}_x)^{1/5}} \tag{4.52}$$

The average drag coefficient for the whole plate is

$$C_{D_{\text{plate}}} = \frac{\text{total drag}}{\text{plate area}} \times \frac{2}{\rho U_\infty^2} = \frac{0.072}{(\mathrm{Re}_l)^{1/5}} \tag{4.53}$$

In our treatment of turbulent boundary layer thickness and drag, we have taken the liberty of ignoring the laminar boundary layer at the beginning of the plate to keep the relations simple. In fact, transition from laminar to turbulent will occur near $\mathrm{Re}_l = 5 \times 10^5$. For the calculation of drag, Prandtl suggested treating the flow as turbulent from the leading edge and then reducing the drag in the region from the leading edge to the transition point to correspond to laminar drag in this region. The following example will provide clarification.

Example 4.17.1

Atmospheric air at 25°C flows past a thin smooth plate that is 2 m wide and 5 m long. The air velocity U_∞ is 10 m/s parallel to the long side of the plate. Assume transition from laminar to turbulent boundary layer flow at a Reynolds number of 5×10^5 (based on length).

(a) At what distance from the plate leading edge does transition occur?
(b) What is the drag force due to the laminar boundary layer from the plate leading edge to the transition point? Include both sides of the plate.
(c) What would be the drag over this same portion of the plate (part b) if we assumed turbulent boundary layer flow?
(d) What is the drag on both sides of the whole plate if we assume turbulent boundary layer flow throughout?
(e) Correct the drag for the whole plate according to Prandtl's suggestion.

Solution:

For atmospheric air at 25°C,

$$\mu = 1.83 \times 10^{-5} \text{ kg/ms}, \qquad \rho = 1.18 \text{ kg/m}^3$$

(a) $$\text{Re}_l = \frac{\rho U_\infty l}{\mu} = 5 \times 10^5$$

$$l(\text{transition}) = \frac{5 \times 10^5 \times (1.83 \times 10^{-5} \text{ kg/ms})}{(1.18 \text{ kg/m}^3)(10 \text{ m/s})} = 0.78 \text{ m}$$

(b) Drag force on both sides of the plate due to laminar flow over the first 0.78 m [use equation (4.29)]:

$$\text{Drag} = 2(\text{sides}) \times 0.655b\sqrt{\mu\rho U_\infty^3 l} \quad (\text{per side})$$

$$= 2 \times 0.655 \times 2 \times (1.83 \times 10^{-5} \times 1.18 \times 10^3 \times 0.78)^{1/2} = 0.34 \text{ N}$$

(c) Drag over first 0.78 m if flow had been turbulent [use equation (4.51)]:

$$\text{Drag} = 2 \times 0.036\rho U_\infty^2\, bl(\text{Re}_l)^{-1/5}$$

$$= 2 \times 0.036 \times 1.18 \times 10^2 \times 2 \times 0.78\,(5 \times 10^5)^{-1/5} = 0.96 \text{ N}$$

(d) For turbulent boundary layer flow over the whole plate,

$$\text{Drag} = 2 \times 0.036 \times 1.18 \times 10^2 \times 2 \times 5\left(\frac{1.83 \times 10^{-5}}{1.18 \times 10 \times 5}\right)^{1/5} = 4.24 \text{ N}$$

(e) Using Prandtl's suggestion,

$$\text{Corrected drag} = (\text{part d}) - (\text{part c}) + (\text{part b})$$

$$= 4.24 - 0.96 + 0.34 = 3.62 \text{ N}$$

PROBLEMS

4.1. Determine the coefficients for a fourth-degree polynomial to be used to represent the velocity profile in a laminar boundary layer. Begin by expressing the velocity ratio u/U_∞ as a fourth-degree polynomial with unknown coefficients. Next, state the boundary conditions to be satisfied, one for each unknown coefficient in the fourth-degree polynomial. *Answer:* $u/U_\infty = 2\eta - 2\eta^3 + \eta^4$

4.2. A fluid with density ρ, viscosity μ, and free-stream velocity U_∞ forms a boundary layer of thickness δ at some location x along a flat plate. Assuming a first-degree (linear) velocity profile for the laminar boundary layer, calculate δ_1/δ, δ_2/δ, and τ_{wx}. *Answer:* $\frac{1}{2}$, $\frac{1}{6}$, uU_∞/δ

4.3. Repeat Problem 4.2 for a velocity profile represented by a second-degree polynomial. *Answer:* $\frac{1}{3}$, $\frac{2}{15}$, $2uU_\infty/\delta$

4.4. Repeat Problem 4.2 for a velocity profile represented by a third-degree polynomial. *Answer:* $\frac{3}{8}$, $\frac{39}{280}$, $\frac{3}{2}\, uU_\infty/\delta$

4.5. Repeat Problem 4.2 for a turbulent boundary layer represented by $u/U_\infty = \eta^{1/7}$. *Answer:* $\frac{1}{8}$, $\frac{7}{72}$, ∞

4.6. Consider a fluid having density ρ flowing over a solid surface through a cross section of height δ and width b.

(a) Calculate the mass and momentum flow rates through the specified area, assuming the fluid velocity is everywhere equal to U_∞. *Answer:* $\rho b\delta U_\infty$, $\rho b\delta U_\infty^2$

(b) Repeat part a using calculus and assuming a second-degree velocity profile ($u/U_\infty = 2\eta - \eta^2$). *Answer:* $\frac{2}{3}\rho b\delta U_\infty$, $\frac{8}{15}\,\rho b\delta U_\infty^2$

(c) Determine the reductions in mass and momentum flow rates through the specified area by subtracting part b from part a.

(d) Using the defining equation for displacement thickness δ_1, calculate δ_1 and the mass-flow deficit and compare with part c. (They should agree.)

(e) Using the momentum thickness δ_2, calculate the momentum deficit and compare with part c. (These answers do not agree. The reduction in momentum flow rate through the specified area has two causes:

(1) The mass flow rate is reduced (δ_1 effect).

(2) The momentum of the actual mass flow rate is itself reduced (δ_2 effect). The total momentum reduction given by part c should be equivalent to free-stream momentum through an area of height $\delta_1 + \delta_2$ and width b.)

4.7. (a) Using the von Kármán momentum integral for boundary layer flow with no pressure gradient, and assuming a second-degree velocity profile, derive an equation for laminar boundary layer thickness at any location x along a flat plate. *Answer:* $\delta = 5.48/\sqrt{\mathrm{Re}_x}$

(b) Water at 25°C flows with $U_\infty = 1$ m/s over a flat plate. Assuming that laminar turbulent transition begins at $\mathrm{Re}_x = 5 \times 10^5$, find the thickness of the laminar boundary layer just before transition. Calculate δ at several points between the plate leading edge and the transition point, and sketch the shape of the developing laminar boundary layer. *Answer:* $\delta_{\text{transition}} = 3.5$ mm

(c) Repeat part b for $U_\infty = 2$ m/s. *Answer:* $\delta_{\text{transition}} = 1.75$ mm

(d) Repeat part b for air at atmospheric pressure and 25°C. In addition, find the friction drag on the plate between the leading edge and transition point, assuming the plate is 1 m wide. *Answer:* $\delta_{\text{transition}} = 0.061$ m; drag = 0.0096 N

4.8. Water at 25°C flows over a flat plate with $U_\infty = 10$ m/s. Find the distance from the leading edge to where $\text{Re}_x = 10^7$, and find the turbulent boundary layer thickness at this point, assuming turbulence right from the leading edge, and the drag for a plate 1 m wide. If the boundary layer had been laminar up to the same point on the plate, what would be the laminar boundary layer thickness and drag? *Answers:* turbulent $\delta = 0.0136$ m, drag = 129 N; laminar $\delta = 0.00136$ m, drag = 19 N

4.9. A boat planes over smooth water at 50 km/hr. The flat bottom area contacting the water at this speed is 1.8 m wide and 2 m long. The frontal area of boat and rider is 1.4 m^2, and the drag coefficient relating to frontal area is 0.5. Estimate the total drag force due to water and air and the power required to propel the boat. *Answer:* drag = 874 N, power = 12.1 kW

Steady-State Conduction of Heat 5

Conduction is one of three modes by which heat is transferred from one part of an object to another; the other two processes are convection and radiation. In an opaque solid, conduction is the only mode of heat transfer that can occur. In gases and liquids, however, flow of heat by conduction is likely to occur simultaneously with flow of heat by convection and radiation.

Conductive heat transfer always occurs spontaneously from a region at a high temperature to one at a lower temperature. The mathematical description of this process is known as Fourier's equation of heat conduction, which was introduced in Chapter 1 (Section 1.4.2).

5.1 Fourier's Heat Conduction Equation

Fourier's equation states that the heat flux is proportional to the negative temperature gradient. Equation (1.8) is the mathematical equivalent of this statement for heat flow in the x direction only.

$$q = -kA\frac{dT}{dx} \tag{1.8}$$

where k is a proportionality constant called the *thermal conductivity* of the material. Most of our discussion in this chapter will be concerned with applications of this basic equation of heat conduction.

5.2 Heat Conduction through a Plane Wall

Let us consider the plane wall of Figure 5.1 with face area A, thickness L, and face temperatures T_1 and T_2, with $T_1 > T_2$. For steady-state conditions, q is constant. Application of equation (1.8) for conduction in the x direction only gives the following result:

$$\frac{q}{A}\int_0^L dx = -\int_{T_1}^{T_2} k\,dT \tag{5.1}$$

or

$$\frac{qL}{A} = -\int_{T_1}^{T_2} k\,dT \tag{5.2}$$

Integration of (5.2) requires that k as a function of T be known. If an average value, $\bar{k}$, is used, where

$$\bar{k} = \frac{1}{T_1 - T_2}\int_{T_1}^{T_2} k\,dT \tag{5.3}$$

equation (5.2) may be written as follows:

$$q = \frac{\bar{k}A(T_1 - T_2)}{L} \tag{5.4}$$

We find it useful to rewrite equation (5.4) in the following way to emphasize the analogy between heat conduction and flow of an electric current.

$$q = \frac{\Delta T}{R} \tag{5.5}$$

where $R = L/\bar{k}A$. $\Delta T = T_1 - T_2$ is interpreted as the potential that causes the heat current q to flow through the thermal resistance R. The electric

Figure 5.1 Plane wall

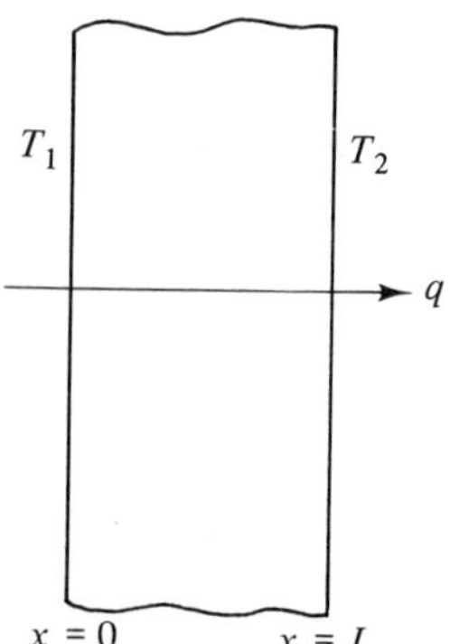

circuit analog of the plane wall can then be represented as follows:

$$q \longrightarrow \qquad T_1 \circ\!\!-\!\!\!-\!\!\!\text{/\/\/\/}\!\!-\!\!\!-\!\!\circ\, T_2 \qquad R = \frac{L}{\bar{k}A}$$

5.3 Heat Conduction through a Composite Wall

Walls of structures such as furnaces or buildings are usually made of several different materials. Such a composite wall made from three different materials placed in layers is illustrated in Figure 5.2. With a fixed temperature drop across the entire wall, conditions are steady and the rate of heat conduction through the wall, q, is constant.

An expression for the heat transfer through the wall of area A is obtained by applying equation (5.4) to each layer.

$$q = \frac{\bar{k}_1 A(T_1 - T_2)}{x_1} = \frac{\bar{k}_2 A(T_2 - T_3)}{x_2} = \frac{\bar{k}_3 A(T_3 - T_4)}{x_3} \tag{5.6}$$

The sum of the temperature differences across each layer must equal the overall temperature difference across the wall.

$$(T_1 - T_2) + (T_2 - T_3) + (T_3 - T_4) = (T_1 - T_4) = \sum \Delta T$$

Substituting from equation (5.6) into this expression gives the following relationship:

$$\sum \Delta T = q\left(\frac{x_1}{\bar{k}_1 A} + \frac{x_2}{\bar{k}_2 A} + \frac{x_3}{\bar{k}_3 A}\right)$$

This may also be written in the form of equation (5.5).

$$q = \frac{\sum \Delta T}{\sum R} \tag{5.7}$$

Figure 5.2 Composite wall

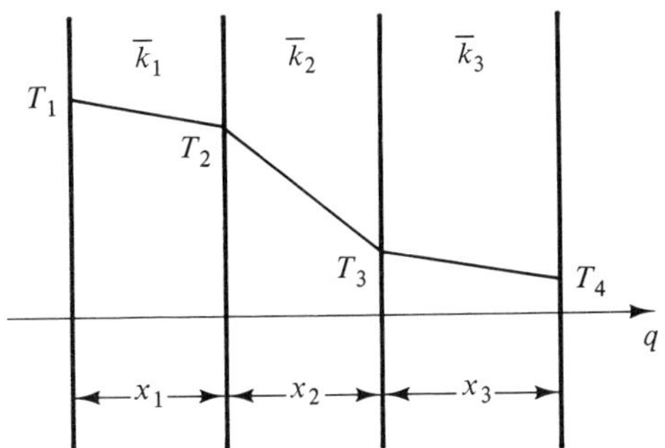

where the total thermal resistance of the wall is

$$\sum R = \frac{x_1}{\bar{k}_1 A} + \frac{x_2}{\bar{k}_2 A} + \frac{x_3}{\bar{k}_3 A}$$

The electric circuit analog of the composite wall thus contains three resistances in series.

$$q \longrightarrow \quad T_1 \quad R_1 = \frac{x_1}{\bar{k}_1 A} \quad T_2 \quad R_2 = \frac{x_2}{\bar{k}_2 A} \quad T_3 \quad R_3 = \frac{x_3}{\bar{k}_3 A} \quad T_4$$

Example 5.3.1

A furnace wall is constructed of the following materials: 7 cm of kaolin firebrick on the inside, followed by 15 cm of kaolin insulating brick and 10 cm of masonry brick. If the inner surface temperature is 670°C and the outer surface temperature is 85°C, estimate the heat flux through the furnace wall.

The average value of the thermal conductivity is obtained from data in Table A.3 (k W/m-°C): kaolin firebrick, $\bar{k} = 0.17$; kaolin insulating, $\bar{k} = 0.26$; masonry, $\bar{k} = 0.66$.

The total thermal resistance of the wall for $A = 1\ \text{m}^2$ is

$$\sum R = \frac{7}{100 \times 0.17} + \frac{15}{100 \times 0.26} + \frac{10}{100 \times 0.66} = 1.14$$

Hence, the heat flux is

$$\frac{q}{A} = \frac{\sum \Delta T}{\sum R} = \frac{670 - 85}{1.14} = 513\ \text{W/m}^2$$

5.4 Surface Coefficient of Heat Transfer

The heat conducted through the wall of a building comes from the warm air on one side of the wall and is passed to the cooler air on the other side. Figure 5.3 is a sketch of the temperature profile between the warm and cool air. Under steady conditions, the rate of heat transfer q is again constant and an expression similar to equation (5.6) can be written.

$$q = \left(\frac{\bar{k}}{x}\right)_h A(T_1 - T_2) = \left(\frac{\bar{k}}{x}\right)_w A(T_2 - T_3) = \left(\frac{\bar{k}}{x}\right)_c A(T_3 - T_4) \tag{5.8}$$

where the subscripts h and c refer to the warm and cool air, respectively, and subscript w refers to the wall.

In the fluids, the distance x_h and x_c are difficult to define experimentally. The ratio (k/x) is therefore given the symbol h and is called the *surface coefficient* or *convection coefficient* of heat transfer. Equation (5.8) is then

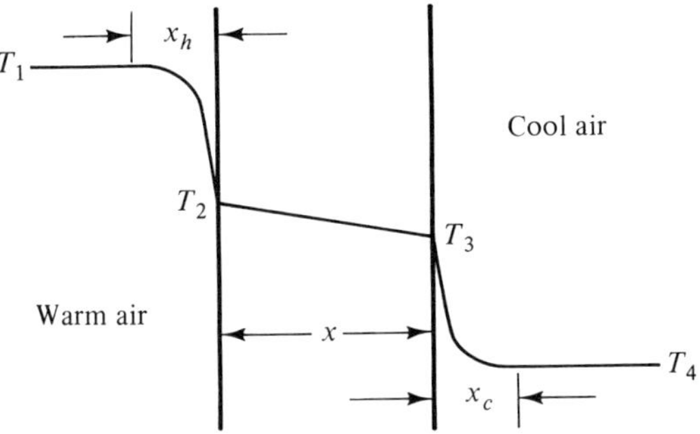

Figure 5.3 Plane wall in contact with air

written as follows:

$$q = h_h A(T_1 - T_2) = \left(\frac{k}{x}\right)_w A(T_2 - T_3) = h_c A(T_3 - T_4) \tag{5.9}$$

Summing the individual temperature differences again gives equation (5.7):

$$q = \frac{\sum \Delta T}{\sum R} \tag{5.7}$$

$$\text{where } \sum \Delta T = T_1 - T_4$$

$$\sum R = \frac{1}{h_h A} + \left(\frac{x}{k}\right)_w \frac{1}{A} + \frac{1}{h_c A}$$

The electric circuit analog is similar to that for the composite wall except that R_1 and R_2 are expressed in terms of the coefficients h_h and h_c.

$q \longrightarrow$ T_1 — T_2 — T_3 — T_4

$$R_1 = \frac{1}{h_h A} \qquad R_2 = \left(\frac{x}{k}\right)_w \frac{1}{A} \qquad R_3 = \frac{1}{h_c A}$$

We may also write equation (5.7) in the following way:

$$q = UA \sum \Delta T \tag{5.10}$$

Comparison with (5.7) indicates that

$$\sum R = \frac{1}{UA}$$

U is commonly referred to as the *overall coefficient of heat transfer*.

We have defined the coefficient h at this point because there are several interesting problems we wish to discuss in which conduction and convection are combined. Convective heat transfer is a subject in itself, however, which we will discuss in more detail in Chapter 6.

5.5 Heat Conduction through a Thick-Walled Tube

There are many examples in industry where heat is conducted through a cylindrical surface. Steam pipes and heat exchanger tubes are two examples of great industrial importance.

Figure 5.4 is a sketch of the cross section of a long tube with the inner surface at a temperature T_1, which is higher than the temperature of the outer surface, which is T_2. The heat is therefore flowing in the direction of the tube radius and through an area A that increases as the radial distance increases.

Since A is not a constant, consider the cylindrical element of radius r and thickness dr. Application of Fourier's equation (1.8) to this element gives

$$q = -kA \frac{dT}{dr} \tag{5.11}$$

$$\text{where } A = 2\pi rL$$

$$L = \text{cylinder length}$$

Thus

$$q = -2\pi krL \frac{dT}{dr}$$

For steady conditions, q is again constant. Separating variables and integrating gives:

$$q \int_{r_1}^{r_2} \frac{dr}{r} = -2\pi \bar{k} L \int_{T_1}^{T_2} dT$$

where an average value of k has been used. Solving for q,

$$q = \frac{2\pi \bar{k} L (T_1 - T_2)}{\ln(r_2/r_1)} \tag{5.12}$$

Figure 5.4 Thick-walled tube

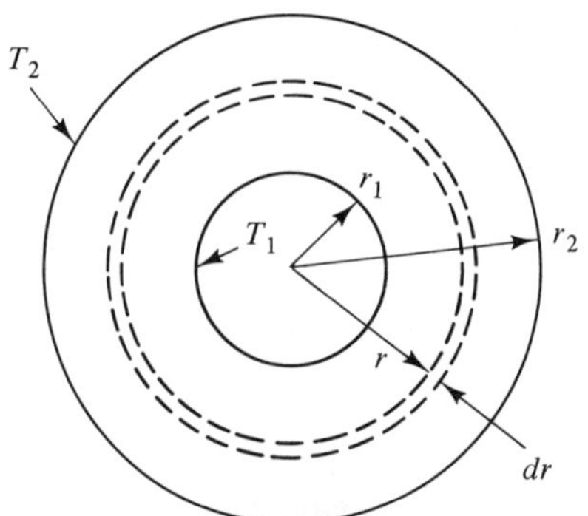

If equation (5.12) is written in the same form as (5.5), that is, as

$$q = \frac{\Delta T}{R} \tag{5.5}$$

the thermal resistance of the cylindrical wall is seen to be

$$R = \frac{1}{2\pi \bar{k} L} \ln \frac{r_2}{r_1}$$

It may be convenient for certain problems to write equation (5.12) in the same form as (5.4), using a mean radius r_m.

$$q = \frac{2\pi \bar{k} r_m L(T_1 - T_2)}{r_2 - r_1} \tag{5.13}$$

Comparisons of equations (5.12) and (5.13) shows that the mean radius is defined as

$$r_m = \frac{r_2 - r_1}{\ln \dfrac{r_2}{r_1}} = \text{log-mean radius}$$

If the cylinder wall is thin, the arithmetic mean radius is an acceptable approximation of the log-mean radius.

$$r_m \approx \frac{r_1 + r_2}{2}$$

Example 5.5.1

A standard 50-mm[1] steel pipe (60.3-mm outside diameter) carrying steam is insulated with a 10-cm-thick layer of glass wool. The outside surface of the pipe has a temperature of 177°C, while the outside surface of the insulation is at 51°C. Determine the heat loss from the pipe expressed as (a) W/m of pipe length, and (b) Btu/hr-ft of pipe length.

From data in Table A.3, an average value for the thermal conductivity of glass wool insulation is

$$\bar{k} = 0.074 \text{ W/m-°C}$$

Using equation (5.12), the heat loss per unit pipe length is

$$\text{(a)}\quad \frac{q}{L} = \frac{2\pi \bar{k}(T_1 - T_2)}{\ln (r_2/r_1)} = \frac{2\pi \times 0.074(177 - 51)}{\ln (6.03 + 2 \times 10)/6.03}$$

$$= 40 \text{ W/m}$$

$$\text{(b)}\quad \frac{q}{L} = 40 \frac{\text{W}}{\text{m}} \times \frac{\text{Btu/hr}}{0.29307 \text{ W}} \times \frac{12 \times 2.54}{100} \frac{\text{m}}{\text{ft}} = 42 \text{ Btu/hr-ft}$$

[1] This is the nominal pipe size. Actual pipe dimensions are given in reference books such as that by Parrish (5) and Perry (6).

5.6 Heat Conduction through a Composite Cylinder

When a pipe such as a steam line is covered with one or more layers of insulating material to reduce heat losses, the heat must flow through several layers in series. Figure 5.5 is a sketch of the cross section of a long tube where a material with an outer radius r_3 covers a tube with an outer radius of r_2. For steady conditions, the rate of heat transfer q is constant and given by an expression similar to equation (5.12) for each cylindrical layer. Thus, if $T_1 > T_2 > T_3$,

$$q = \frac{2\pi\bar{k}_1 L(T_1 - T_2)}{\ln (r_2/r_1)} = \frac{2\pi\bar{k}_2 L(T_2 - T_3)}{\ln (r_3/r_2)} \tag{5.14}$$

Since the overall temperature difference is just the sum of the temperature drop across each layer, equation (5.14) may be written in the same form as equation (5.7).

$$q = \frac{\sum \Delta T}{\sum R} \tag{5.7}$$

where $\sum \Delta T = T_1 - T_3$

$$\sum R = \frac{1}{2\pi L}\left(\frac{1}{\bar{k}_1} \ln \frac{r_2}{r_1} + \frac{1}{\bar{k}_2} \ln \frac{r_3}{r_2}\right)$$

The electric circuit analog is similar to that for the composite wall.

$q \longrightarrow$ T_1 —R₁— T_2 —R₂— T_3

$$R_1 = \frac{1}{2\pi\bar{k}_1 L}\, ln\, \frac{r_2}{r_1} \qquad R_2 = \frac{1}{2\pi\bar{k}_2 L}\, ln\, \frac{r_3}{r_2}$$

Figure 5.5 Composite cylinder

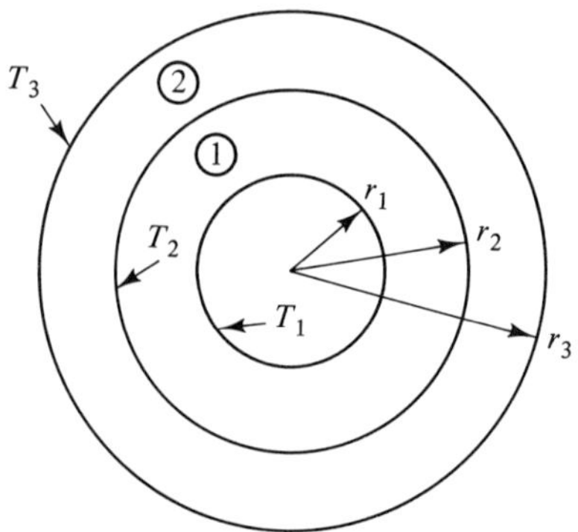

Example 5.6.1

The steam line in Example 5.5.1 is covered with a second layer of asbestos insulation that is 7 cm thick. If the outer surface of the asbestos is at a temperature of 43°C, estimate (a) the new heat loss from the line, and (b) the temperature at the interface between the two layers of insulation.

From Table A.3, the thermal conductivity of asbestos is estimated to be $\bar{k}$ = 0.17 W/m-°C. The total thermal resistance of the two layers of insulation is, per unit of pipe length,

$$\sum R = \frac{1}{2\pi}\left(\frac{1}{\bar{k}_1}\ln\frac{r_2}{r_1} + \frac{1}{\bar{k}_2}\ln\frac{r_3}{r_2}\right)$$

$$= \frac{1}{2\pi}\left(\frac{1}{0.074}\ln\frac{26.03}{6.03} + \frac{1}{0.17}\ln\frac{26.03 + 2 \times 7}{26.03}\right)$$

$$= 3.55$$

Thus, using equation (5.7),

(a) $\dfrac{q}{L} = \dfrac{177 - 43}{3.55} = 38$ W/m of pipe length

which is a reduction in heat loss of only about 5%.

To find the interface temperature, equate the heat flow through the first layer of insulation to that through both layers in series.

(b) $q = \dfrac{T_1 - T_2}{R_1} = \dfrac{T_1 - T_3}{R_1 + R_2}$

Solving for T_2,

$$T_2 = T_1 - \frac{R_1(T_1 - T_3)}{R_1 + R_2} = 177 - \frac{3.15\ (177 - 43)}{3.55} = 58°\text{C}$$

5.7 Heat Conduction through a Thick-Walled Sphere

Conduction through spherical shapes does not occur as frequently in industry as through walls and cylinders. Applications of equation (5.11) with

$$A = 4\pi r^2$$

lead to the following expression for q:

$$q = \frac{4\pi\bar{k}r_1r_2(T_1 - T_2)}{r_2 - r_1} \tag{5.15}$$

The student should verify this derivation.

5.8 Heat Loss from Insulated Pipes

An interesting industrial application of heat conduction through cylindrical walls is the problem of insulating pipe lines such as steam lines and lines carrying refrigerated fluids. Analysis of this problem will show that the addition of insulating material does not always reduce the rate of heat transfer between the pipe and its surroundings.

Suppose a pipe of outer radius r_2 has a layer of insulating material of radius r added to it, as indicated in Figure 5.6. All the thermal resistance is assumed to be in the layer of insulation and in the air that surrounds the entire structure. Expressing this resistance per unit length of pipe gives the following equation for the total thermal resistance:

$$\Sigma R = \frac{1}{2\pi\overline{k}} \ln \frac{r}{r_2} + \frac{1}{2\pi r h} \tag{5.16}$$

where $\overline{k}$ = mean thermal conductivity of the insulation

h = coefficient of heat transfer to the air

Since one term in equation (5.16) increases while the other term decreases as r increases, one would expect ΣR to have a minimum value at some value of r.

If h is assumed to be independent of r, the minimum for the function (5.16) is readily found by calculating the first differential of ΣR with respect to r and putting this equal to zero.

$$\frac{d\,\Sigma R}{dr} = \frac{1}{2\pi\overline{k}r} - \frac{1}{2\pi r^2 h} = 0$$

Solving this expression for r gives the value of the critical radius r_c, which is the radius at which ΣR has its minimum value.

$$r_c = \frac{\overline{k}}{h} \tag{5.17}$$

Figure 5.6 Insulated pipe

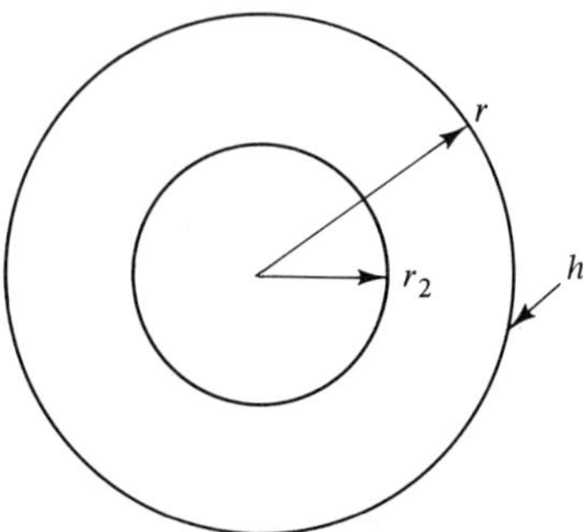

The significance of r_c is that insulation added to the pipe up to this value of r has caused the rate of heat transfer q to increase rather than to decrease as desired. However, if an insulating material is chosen with a $\bar{k}$ value such that r_c is smaller than the outer radius of the pipe, then addition of this material to the pipe surface will always result in a decrease in the rate of heat transfer q.

A more precise analysis of this problem would include the coefficient of heat transfer of the fluid inside the pipe, h_i, and the thermal resistance of the pipe wall

$$R_{\text{wall}} = \frac{1}{2\pi \bar{k}_w} \ln \frac{r_2}{r_1}$$

where r_1 is the inside pipe radius, and allow for the convection coefficient to the air to change as the radius r increases. For instance, according to McAdams (4), for a horizontal pipe,

$$h = 0.27\,(\Delta T / D_0)^{0.25} \tag{5.18}$$

where ΔT = difference in temperature between the surface of the insulation and the bulk air

D_0 = $2r$

Combining all four terms that make up the total resistance, ΣR, results in the following expression for q:

$$\frac{q}{L} = \frac{\Sigma\,\Delta T}{\dfrac{1}{2\pi r_1 h_i} + \dfrac{1}{2\pi \bar{k}_w} \ln \dfrac{r_2}{r_1} + \dfrac{1}{2\pi \bar{k}} \ln \dfrac{r}{r_2} + \dfrac{(2r)^{0.25}}{2\pi r \times 0.27 \Delta T^{0.25}}} \tag{5.19}$$

where $\Sigma\Delta T$ = difference in temperature between the fluid inside the pipe and the bulk air

For a given set of the constants in equation (5.19), a plot of q/L against r can be prepared. The curve will have a maximum that can be used to define r_c. The solution is to choose a value for $\bar{k}$ so that r_c is smaller than the pipe radius r_2. Notice, too, that since ΔT decreases as r increases the solution involves a trial-and-error calculation to determine the outside surface temperature.

5.9 Conduction in Walls with Heat Generation

Heat can be generated within a solid object in a number of ways, such as by passage of an electric current or by a nuclear or chemical reaction. If this heat is to be removed, it must first be transferred by conduction to the surface boundaries of the object.

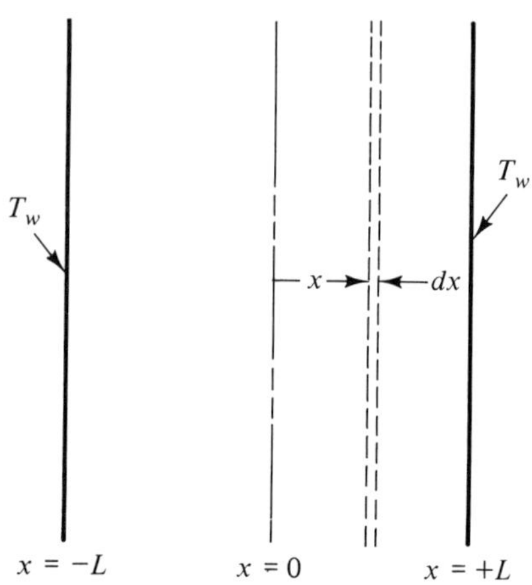

Figure 5.7 Plane wall with heat generation

Let us consider a plane wall of thickness $2L$ with the two surfaces at the same temperature T_w, as illustrated in Figure 5.7. Let $\dot{q}$ be the rate of heat generation per unit wall volume, assumed to be uniform throughout the wall.

For the heat to be conducted to the two wall surfaces a temperature gradient must be established within the wall. This temperature profile will be symmetrical in the present case, with a maximum at the center of the wall. One-half of the total heat generated will pass through each surface.

To find this temperature profile, we will consider an element of the wall of thickness dx at a distance x from the center. For steady-state conditions, a heat balance on the element can be written as follows:

$$\text{Heat entering} + \text{heat generated} = \text{heat leaving} \tag{5.20}$$

The first term of equation (5.20) is given by Fourier's equation (1.8).

$$q\ (\text{in}) = -\bar{k}A\,\frac{dT}{dx}$$

where $\bar{k}$ and A are the mean thermal conductivity and area of the wall, respectively.

The second term of the balance is the product of $\dot{q}$ and the volume of the element.

$$q\ (\text{gen}) = \dot{q}A\ dx$$

The third term is also obtained by applying Fourier's equation to the conditions at distance $(x + dx)$.

$$\begin{aligned} q\ (\text{out}) &= -\bar{k}A\left[\frac{dT}{dx} + \frac{d}{dx}\left(\frac{dT}{dx}\right)dx\right] \\ &= -\bar{k}A\left(\frac{dT}{dx} + \frac{d^2T}{dx^2}\,dx\right) \end{aligned}$$

Substituting for the three terms in equation (5.20) and simplifying gives the following differential equation:

$$\frac{d^2T}{dx^2} + \frac{\dot{q}}{\overline{k}} = 0 \tag{5.21}$$

The solution of equation (5.21) is readily obtained by two integrations remembering that $\dot{q}$ and $\overline{k}$ are constants. The first integration results in

$$\frac{dT}{dx} = -\left(\frac{\dot{q}}{\overline{k}}\right)x + C_1 \tag{5.22}$$

and the second gives the result

$$T = -\left(\frac{\dot{q}}{\overline{k}}\right)\frac{x^2}{2} + C_1x + C_2 \tag{5.23}$$

where C_1 and C_2 are constants of integration.

Two boundary conditions are required to evaluate C_1 and C_2. These conditions are the following:

$$x = 0, \frac{dT}{dx} = 0 \quad \text{or} \quad T = T^0$$

$$x = \pm L, T = T_w$$

From the first of these conditions and equation (5.22), the result is

$$C_1 = 0$$

From either of the other conditions and equation (5.23),

$$C_2 = T^0 \quad \text{or} \quad C_2 = T_w + \left(\frac{\dot{q}}{\overline{k}}\right)\frac{L^2}{2}$$

Substituting for C_1 and C_2 into (5.23) gives the solution to (5.21). This can be expressed in a number of ways, all of which are equivalent mathematical expressions for the temperature profile in the wall.

$$T = T^0 - \left(\frac{\dot{q}}{\overline{k}}\right)\frac{x^2}{2} \tag{5.24}$$

$$T = T_w + \frac{\dot{q}}{2\overline{k}}(L^2 - x^2) \tag{5.25}$$

$$\frac{T - T^0}{T_w - T^0} = \left(\frac{x}{L}\right)^2 \tag{5.26}$$

If the two surface temperatures of the wall are not the same, the problem is no longer symmetrical. More heat will pass through one surface than the other, and the maximum in the temperature profile will not occur at the center of the wall. Let

$$T_{w1} = \text{surface temperature at } x = -L$$

$$T_{w2} = \text{surface temnerature at } x = L$$

Substituting these two boundary conditions into equation (5.23) gives

$$T_{w1} = -\left(\frac{\dot{q}}{k}\right)\frac{L^2}{2} - C_1 L + C_2$$

$$T_{w2} = -\left(\frac{\dot{q}}{k}\right)\frac{L^2}{2} + C_1 L + C_2$$

Solving for C_1 and C_2 gives

$$C_1 = \frac{T_{w2} - T_{w1}}{2L}, \qquad C_2 = \frac{1}{2}\left(T_{w1} + T_{w2} + \frac{\dot{q}L^2}{k}\right) \tag{5.27}$$

The general solution is thus obtained by substituting equation (5.27) into (5.23).

$$T = \frac{1}{2}\left[\frac{\dot{q}}{k}(L^2 - x^2) + \left(1 - \frac{x}{L}\right)T_{w1} + \left(1 + \frac{x}{L}\right)T_{w2}\right] \tag{5.28}$$

The position of the maximum in the temperature profile is found by putting

$$\frac{dT}{dx} = 0$$

Thus

$$\frac{dT}{dx} = \frac{1}{2}\left[\frac{\dot{q}}{k}(-2x) - \frac{T_{w1}}{L} + \frac{T_{w2}}{L}\right] = 0$$

Solving for x gives the distance from the center, x_0, where the temperature has its maximum value T^0.

$$x_0 = \frac{k}{2\dot{q}L}(T_{w2} - T_{w1}) \tag{5.29}$$

Substituting x_0 for x in equation (5.28) gives the value of the maximum temperature T^0.

$$T^0 = \frac{1}{2}\left[\frac{\dot{q}L^2}{k} + T_{w1} + T_{w2} + (T_{w1} - T_{w2})^2\frac{k}{4\dot{q}L^2}\right] \tag{5.30}$$

5.10 Convection at Walls with Heat Generation

For the plane wall illustrated in Figure 5.7, suppose each surface is in contact with fluid at a bulk temperature of T_∞. The steady-state temperature profile in the wall and fluid is again symmetrical, as illustrated in Figure 5.8. All the heat generated within the wall passes into the fluid at the two surfaces under the temperature potential $(T_w - T_\infty)$. This may be stated mathematically for steady conditions as follows:

$$\begin{aligned} q\ (\text{gen}) &= \dot{q}(2AL) \\ &= 2hA(T_w - T_\infty) \end{aligned} \tag{5.31}$$

where h is the coefficient of heat transfer at the wall surface of area A. Solving for the surface temperature T_w gives

$$T_w = T_\infty + \frac{\dot{q}L}{h} \tag{5.32}$$

Substituting for T_w from equation (5.32) in (5.25),

$$T = T_\infty + \frac{\dot{q}L}{h} + \frac{\dot{q}}{2k}(L^2 - x^2) \tag{5.33}$$

which gives the temperature at any distance x in the wall.

If the maximum temperature in the wall T^0 at $x = 0$ is desired, then, from (5.33),

$$T^0 = T_\infty + \dot{q}L\left(\frac{1}{h} + \frac{L}{2k}\right) \tag{5.34}$$

Figure 5.8 Wall with heat generation and surface convection

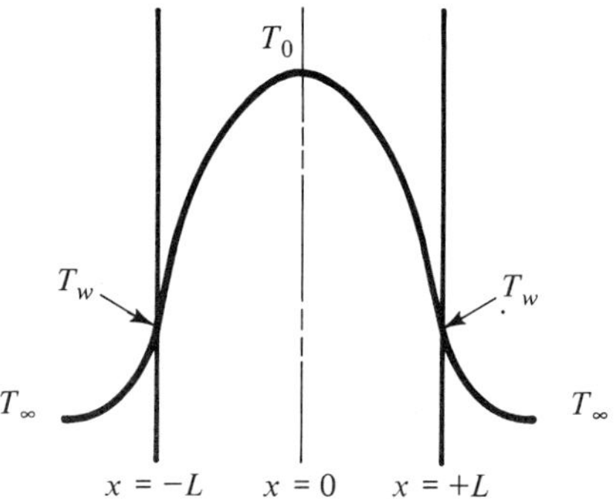

Example 5.10.1

A large plane wall 15 cm thick has heat generated internally at the rate of 155,000 W/m³. Both sides of the wall are exposed to the environment at 25°C. The heat transfer coefficient at the wall surface is estimated to be 280 W/m²-°C. The thermal conductivity of the wall is 21 W/m-°C. Estimate the maximum temperature in the wall.

The maximum temperature occurs at the center of the wall in this case. This can be calculated using equation (5.25), provided the wall temperature T_w is known. This in turn is found from equation (5.32), which is derived from (5.31), which is a statement that all the heat generated within the wall must pass out through the two surfaces of area A.

From (5.32),

$$T_w = T_\infty + \frac{\dot{q}L}{h} = 25 + \frac{155{,}000 \times 15}{280 \times 100 \times 2} = 66.5°\text{C}$$

From equation (5.25) for $x = 0$,

$$T^0 = T_w + \frac{\dot{q}L^2}{2k} = 66.5 + \frac{155{,}000\ (15/2)^2}{2 \times 21\ (100)^2} = 87.5°\text{C}$$

5.11 Conduction in Cylinders with Heat Generation

Wires carrying an electric current and nuclear reactor fuel rods are examples of cylindrical objects where heat is generated internally. In such structures, it is important that this heat be removed efficiently in order that melting and physical damage to the wire or rod do not occur.

Figure 5.9 represents the cross section of a long rod in which heat is being generated. This heat must flow by conduction to the surface of the cylinder to be removed, and thus a radial temperature gradient must exist in the rod.

Figure 5.9 Cylinder with internal heat generation

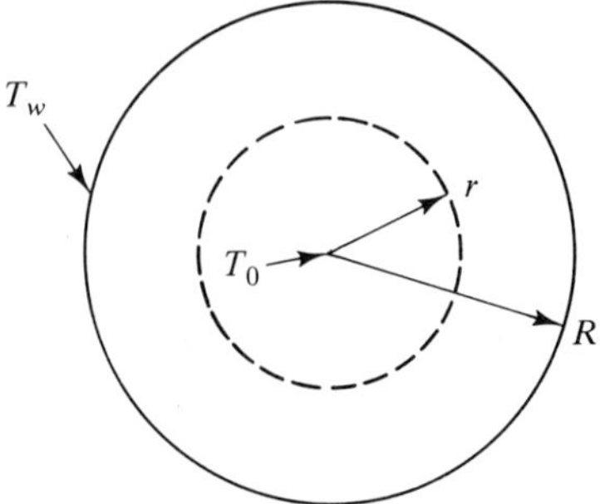

An expression for calculating this temperature gradient is obtained as follows. Consider an element of the rod of radius r. For steady-state conditions, all the heat generated within the element must pass through the cylindrical surface of radius r by conduction. Stated mathematically,

$$\dot{q}(\pi r^2 L) = -k(2\pi r L)\frac{dT}{dr}$$

or, simplifying,

$$\frac{dT}{dr} = -\frac{\dot{q}}{2k}r \tag{5.35}$$

Integrating equation (5.35) gives

$$T = -\frac{\dot{q}}{4k}r^2 + C_1 \tag{5.36}$$

The boundary condition $T = T_0$ for $r = 0$ gives $C_1 = T_0$, from which we obtain

$$T_0 - T = \frac{\dot{q}}{4k}r^2 \tag{5.37}$$

Alternatively, the boundary condition $T = T_w$ for $r = R$ gives

$$C_1 = T_w + \frac{\dot{q}}{4k}R^2$$

so that

$$T - T_w = \frac{\dot{q}}{4k}(R^2 - r^2) \tag{5.38}$$

Another way of expressing the temperature profile is to note from equation (5.37) that

$$T_0 - T_w = \frac{\dot{q}R^2}{4k} \tag{5.39}$$

Dividing equation (5.38) by (5.39) gives

$$\frac{T - T_w}{T_0 - T_w} = 1 - \left(\frac{r}{R}\right)^2 \tag{5.40}$$

Equations (5.37), (5.38), and (5.40) are different ways of expressing the same temperature profile in the rod.

Example 5.11.1

A stainless steel wire 3.43 mm in diameter and 1 m long has a dc current of 100 amperes passing through it. If the wire is immersed in water at the normal boiling point, estimate the maximum temperature in the wire. Take the heat-transfer coefficient at the wire surface, $h = 5600$ W/m²-°C, the thermal conductivity of stainless steel, $k = 22$ W/m-°C, and the resistivity of steel as 70 microohms (μΩ)-cm.

The surface temperature of the wire can be found by using the convection equation together with the fact that all the heat generated within the wire must pass through this surface. Thus

$$q = \dot{q}(AL) = hA_s(T_w - T_\infty) = I^2R_w$$

where A and A_s are the cross-sectional and surface areas of the wire, respectively. The wire resistance R_w is given by

$$R_w = \frac{\rho L}{A} = \frac{70 \times 10^{-6} \times 100}{(\pi/4) \times (0.343)^2} = 0.0758\ \Omega$$

Hence

$$T_w = T_\infty + \frac{I^2R_w}{hA_s} = 100 + \frac{(100)^2\ 0.0758}{5600\ \pi \times (3.43/1000) \times 1}$$

$$= 112.6°\text{C}$$

Equation (5.39) can now be used to estimate the maximum temperature in the wire.

$$T^0 = T_w + \frac{\dot{q}R^2}{4k} = 112.6 + \frac{(100)^2\ 0.0758}{\frac{\pi}{4}\left(\frac{3.43}{1000}\right)^2 \times 1} \times \left(\frac{3.43}{2 \times 1000}\right)^2 \times \frac{1}{4 \times 22}$$

$$= 112.6 + 2.7 \approx 115°\text{C}$$

5.12 Heat Transfer from Fins

There are many instances where a tube or a plate is not able to transfer as much heat as is desired. A common design remedy is to extend the surface in some way so that the rate of heat transfer is increased. The radiator of an automobile, for instance, has thin strips of metal bonded to the radiator tubes so that the cooling of the radiator fluid is increased.

A simple type of extended surface is a rod welded to a base plate, as illustrated in Figure 5.10. Heat flows from the plate and along the rod by conduction. At the same time heat is lost from the surface of the rod by convection to the surrounding fluid.

We will analyze this simple fin in detail to learn how to assess the effectiveness of such devices. Consider an element of the rod of length dx located

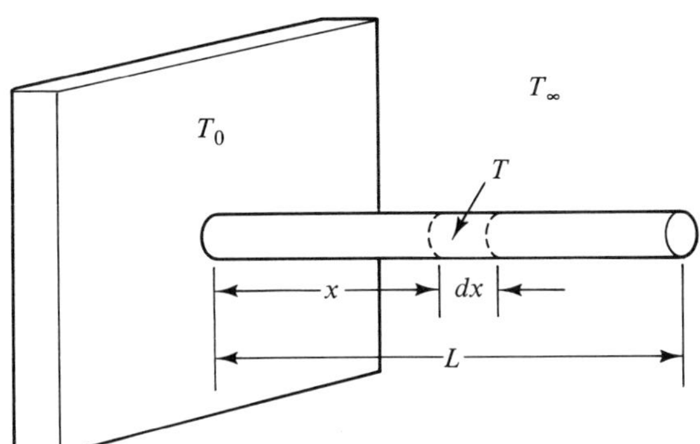

Figure 5.10 Rod-type extended surface

at a distance x from the plate. If radial temperature gradients are assumed to be small and are neglected, the temperature of the rod element may be taken as T. For steady-state conditions, a heat balance for the element is written as follows:

$$\begin{pmatrix}\text{Heat into element}\\ \text{by conduction}\end{pmatrix} = \begin{pmatrix}\text{heat out of element}\\ \text{by conduction}\end{pmatrix} + \begin{pmatrix}\text{heat out of element}\\ \text{by convection}\end{pmatrix} \tag{5.41}$$

The first term of this balance is obtained by applying Fourier's equation to conditions at x:

$$q\ (\text{in}) = -\bar{k}A\left.\frac{dT}{dx}\right|_x$$

while the second term is obtained by applying the same equation to conditions at $x + dx$.

$$q\ (\text{out}) = -\bar{k}A\left.\frac{dT}{dx}\right|_{x+dx} = -\bar{k}A\left[\frac{dT}{dx} + \frac{d}{dx}\left(\frac{dT}{dx}\right)dx\right]$$

$$= -\bar{k}A\left(\frac{dT}{dx} + \frac{d^2T}{dx^2}dx\right)$$

where A = cross-sectional area of the rod

The third term of equation (5.41) is written for convective heat transfer at the rod surface dA_s.

$$q\ (\text{out by convection}) = h\,dA_s(T - T_\infty)$$

$$= hP\,dx\,(T - T_\infty)$$

where P = perimeter of the rod

T_∞ = temperature of the fluid at some distance from the rod.

Substituting for each term in equation (5.41) and simplifying gives the following differential equation:

$$\frac{d^2T}{dx^2} - \frac{hP}{kA}(T - T_\infty) = 0 \tag{5.42}$$

We may write this equation in a simpler form if the following substitutions are made.

$$\theta = T - T_\infty, \qquad m^2 = \frac{hP}{kA}$$

We thus obtain

$$\frac{d^2\theta}{dx^2} - m^2\theta = 0 \tag{5.43}$$

Equation (5.43) is a second-order differential equation with constant coefficients for which the general solution is

$$\theta = C_1e^{-mx} + C_2e^{mx} \tag{5.44}$$

where C_1 and C_2 are constants of integration. Two boundary conditions are required to evaluate these constants. Depending on what we choose for these boundary conditions, we can speak of different types of fins.

5.12.1 The Long Rod Fin. A *long* fin is one whose temperature changes from T_0 at the base to the environment temperature T_∞ at the fin tip. The two boundary conditions required to evaluate the constants in equation (5.44) are

$$\text{For } x = 0, \qquad \theta = T_0 - T_\infty = \theta_0 \tag{5.45}$$

$$\text{For } x = \infty, \qquad \theta = T_\infty - T_\infty = 0 \tag{5.46}$$

Thus, from equations (5.45) and (5.44), we obtain

$$\theta_0 = C_1 + C_2 \tag{5.47}$$

and from equations (5.46) and (5.44), we obtain

$$0 = C_1(0) + C_2(\infty) \tag{5.48}$$

Equation (5.48) is valid if $C_2 = 0$. Using this result, we obtain from (5.47) that $C_1 = \theta_0$. Substitution for C_1 and C_2 in (5.44) then gives the following expression for the temperature distribution along the length of the fin.

$$\theta = \theta_0e^{-mx} \tag{5.49}$$

We must remember, however, that (5.49) is not generally valid, but only applies to those fins whose tip temperature is equal to or reasonably close to the temperature of the surrounding bulk fluid.

5.12.2 Fin with Insulated Tip. Another type of fin we can discuss is one whose tip is insulated so that no heat transfer occurs at this point. The boundary condition in this instance may be expressed as

$$\text{For } x = L, \qquad d\theta/dx = 0$$

When combined with boundary condition (5.45) and the constants C_1 and C_2 in equation (5.44) evaluated, the expression for the temperature distribution along the fin is

$$\frac{\theta}{\theta_0} = \frac{\cosh[m(L - x)]}{\cosh mL} \tag{5.50}$$

That this is a different temperature distribution from that for the long fin is readily seen by comparing equations (5.49) and (5.50).

The characteristics of these two fins are further illustrated by the numerical calculations in the following example.

Example 5.12.1

An aluminum rod (k = 228 W/m-°C) 12 mm in diameter and 30 cm long has one end maintained at a temperature of 120°C. The rod extends into air at 20°C and h = 9.0 W/m²-°C. Find:

(a) The temperature distribution along the length of the rod, assuming the rod behaves as a long fin.
(b) The calculated temperature at the tip of the rod, assuming the temperature distribution to be that of a long fin.
(c) The rod length necessary for the tip temperature to be reasonably close to the temperature of the surroundings.
(d) The temperature at the tip of the rod if the tip is well insulated.

For a long fin, the temperature distribution along the fin length is given by equation (5.49).

$$m = \left(\frac{hP}{kA}\right)^{1/2} = \left[\frac{9.0 \times \pi \times 12 \times 10^{-3}}{228 \times (\pi/4)\,(12 \times 10^{-3})^2}\right]^{1/2} = 3.63$$

$$\theta_0 = 120 - 20 = 100°\text{C}$$

(a) Temperature distribution is $\theta = 100e^{-3.63x}$.
(b) For $x = 0.30$ m, $\theta = 100e^{-3.63 \times 0.30} = 34°\text{C}$. Hence the tip temperature is 34 + 20 = 54°C. Equation (5.49) assumes the tip temperature is $T_\infty = 20°\text{C}$. It is therefore concluded that the rod is not acting as a long fin and that (5.49) is not valid for this fin.
(c) If a tip temperature of 25°C is reasonably close to that expected for a long fin, the rod length required is calculated from equation (5.49). Thus $5 = 100e^{-3.63x}$ and $x = \ln(5/100)/-3.63 = 0.83$ m.
(d) If the tip is well insulated, the temperature distribution is given by equation (5.50). Thus

$$\theta = \frac{100 \cosh 3.63(0.30 - 0.30)}{\cosh 3.63 \times 0.30} = 60°\text{C}$$

and the tip temperature is 60 + 20 = 80°C.

5.13 Rate of Heat Transfer from a Fin

A quantity of interest in fin design is the rate at which heat is transferred to the surrounding fluid. This may be calculated in two ways, either (1) by noting that the heat dissipated by the fin must have entered the fin at its base or (2) by evaluating the total convective rate of heat transfer from the surface of the fin. We will illustrate these two methods of calculation for the two types of fin defined in Section 5.12.

5.13.1 Heat Transfer Rate from a Long Fin. Since all the heat dissipated by a fin must have entered the fin by conduction at its base, we can define the total heat transfer rate q in the following way:

$$q = -\bar{k}A \left.\frac{dT}{dx}\right|_{x=0} \tag{5.51}$$

For a long fin, the temperature gradient at $x = 0$ is evaluated using equation (5.49).

$$\frac{dT}{dx} = \frac{d\theta}{dx} = -m\theta_0 e^{-mx}$$

and at $x = 0$

$$\left.\frac{dT}{dx}\right|_{x=0} = -m\theta_0$$

Substitution of this result into equation (5.51) gives the total rate of heat dissipation by the fin.

$$q = \bar{k}Am\theta_0$$

or, since

$$m^2 = \frac{hP}{kA}$$

then

$$q = (hP\bar{k}A)^{1/2}\theta_0 \tag{5.52}$$

A second way to obtain q is to evaluate the convective heat transfer rate over the entire surface of the fin. For the rod element illustrated in Figure 5.10, we express the convection heat transfer rate as

$$\text{Convective rate} = hP(T - T_\infty)\,dx = hP\theta\,dx$$

For the long fin, equation (5.49) gives θ in terms of x. Making this substitution and integrating over the total surface of the rod gives the total heat transfer rate.

$$q = \int_0^\infty hP\theta_0 e^{-mx}\,dx = hP\theta_0\left(-\frac{1}{m}\right)e^{-mx}\Big|_0^\infty$$

Evaluating the integral and substituting for m gives the same equation for q we obtained before, that is, (5.52).

While the rod is not a common type of extended surface used in industry, it does lend itself readily to theoretical treatment and illustration of the basic principles involved in fin design. A more practical type of design is the straight rectangular fin, which is simply a thin plate of metal welded to either the outside wall of a tube or to a base plate. The theoretical treatment of this type of fin is exactly the same as for the rod, so equations such as (5.49), (5.50), and (5.52) also apply to the straight rectangular fin.

5.13.2 Heat Transfer Rate from a Fin with Insulated Tip. For the fin with an insulated tip, the temperature gradient in equation (5.51) must be evaluated using (5.50).

Thus

$$\frac{dT}{dx} = \frac{d\theta}{dx} = \frac{d}{dx}\left|\frac{\theta_0 \cosh[m(L - x)]}{\cosh mL}\right| = \frac{-\theta_0 m \sinh[m(L - x)]}{\cosh mL}$$

and at $x = 0$

$$\left.\frac{dT}{dx}\right|_{x=0} = -\theta_0 m \tanh mL$$

Substitution into equation (5.51) gives, for the rate of heat transfer,

$$q = -\bar{k}A(-m)\theta_0 \tanh mL$$

or

$$q = (hPkA)^{1/2}\, \theta_0 \tanh mL \tag{5.53}$$

Alternatively, evaluating the rate of heat convection from the fin surface gives

$$\text{Convective rate} = hP\theta\, dx = \frac{hP\theta_0 \cosh\,[m(L - x)]}{\cosh mL}\, dx$$

Integrating over the fin length L results in

$$q = \frac{hP\theta_0}{\cosh mL}\int_0^L \cosh\,[m(L - x)]\, dx = (hP\bar{k}A)^{1/2}\, \theta_0 \tanh mL$$

which is equation (5.53).

Example 5.13.1

If the rod in Example 5.12.1 has an insulated tip, determine the rate of heat transfer to the surrounding air. q is given by equation (5.53). Thus

$$q = [9.0 \times \pi \times 0.012 \times 228 \times \frac{\pi}{4}(0.012)^2]^{1/2}\, 100 \tanh(3.63 \times 0.30)$$

$$= 7.5 \text{ W}$$

5.14 Fin Efficiency

The effectiveness of a fin in transferring heat can be represented by an efficiency η. This is defined as the ratio of the actual heat transferred to that which would be transferred if the entire fin were at the base temperature T_0.

$$\eta = \frac{\text{actual heat transferred}}{\text{heat transferred if entire fin is at the base temperature } T_0}$$

We may then write the heat transfer rate q for a fin as follows:

$$q = \eta h A_f \theta_0 \tag{5.54}$$

where A_f = surface area of the fin.

The value of η for a *long* fin with its tip cooled to the environment temperature T_∞ is readily calculated using equation (5.52). Thus

$$\eta = \frac{(hP\bar{k}A)^{1/2}\theta_0}{hPL\theta_0} = \left(\frac{\bar{k}A}{hP}\right)^{1/2}\frac{1}{L} = \frac{1}{mL} \tag{5.55}$$

For the fin with an insulated tip, q is given by equation (5.53).

$$q = (hP\bar{k}A)^{1/2}\theta_0 \tanh mL \tag{5.53}$$

The value of η for this type of fin is therefore given by

$$\eta = \frac{(hP\bar{k}A)^{1/2}\theta_0 \tanh mL}{hPL\theta_0} = \frac{\tanh mL}{mL} \tag{5.56}$$

The majority of fins used in practice, however, are neither long in the sense used previously nor have an insulated tip. They are finite in length and have a significant amount of heat transfer at the fin tip. The efficiency of such fins is not readily determined analytically. Instead, an empirical method is used, the essence of which is to extend the length L by one-half the fin thickness t to approximately correct for the heat transfer at the tip. This method was first proposed by Harper and Brown (2) and has been discussed by several authors (1, 3, 7, 8). Thus L in equation (5.56) is replaced by a corrected length L_c defined by

$$L_c = L + \frac{t}{2} \tag{5.57}$$

An alternative to equations (5.56) and (5.57) is to use plots such as those in Figures 5.11 and 5.12, where

$$L_c = \begin{cases} L + \dfrac{t}{2}, & \text{rectangular fin} \\ L, & \text{triangular fin} \end{cases}$$

$$A_m = \begin{cases} tL_c, & \text{rectangular fin} \\ \dfrac{t}{2}L, & \text{triangular fin} \end{cases}$$

in Figure 5.11 and

$$L_c = L + \frac{t}{2}$$

$$r_{2c} = r_1 + L_c$$

$$A_m = t(r_{2c} - r_1)$$

in Figure 5.12.

Figure 5.11 Efficiencies of triangular and rectangular fins

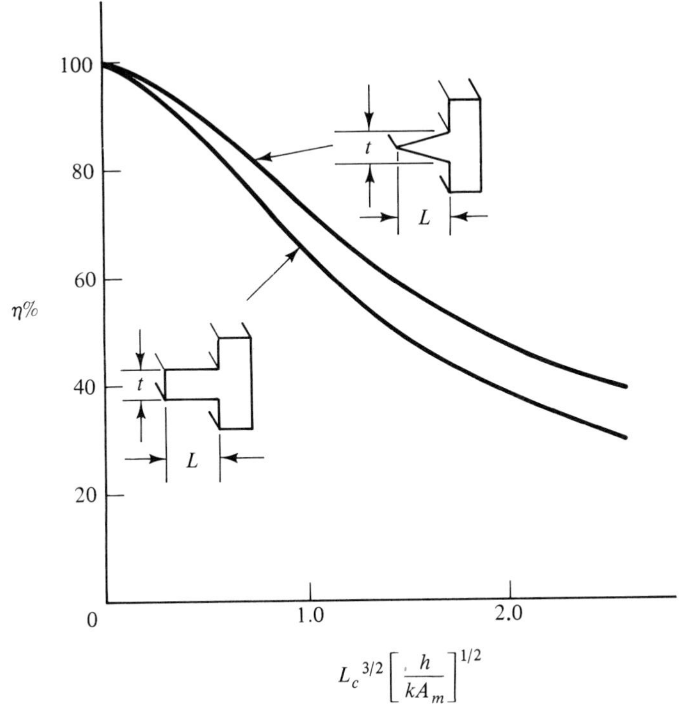

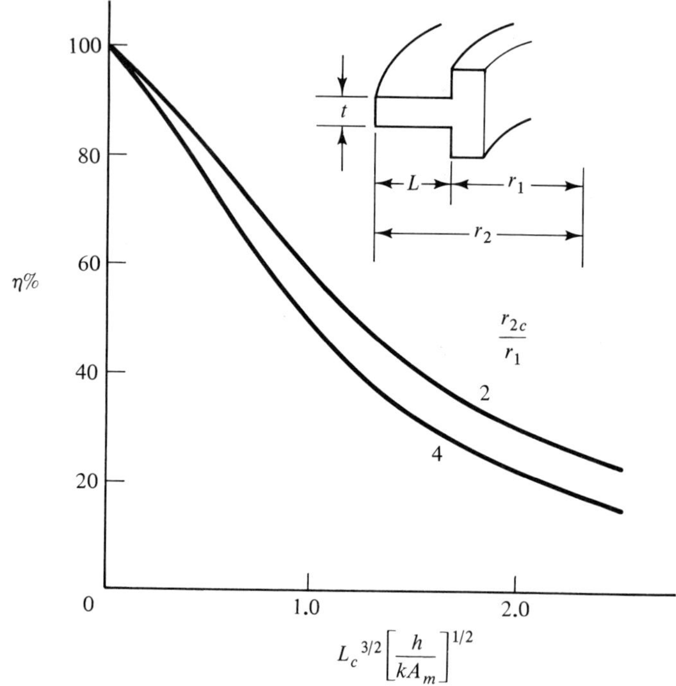

Figure 5.12 Efficiencies of annular fins with rectangular profile

Example 5.14.1

A long aluminum plate fin is attached to the outer surface of a long pipe to enhance heat transfer to the surroundings. The fin is 2 mm thick and extends 90 mm radially outward from the pipe. The pipe wall temperature is 220°C, the surrounding air temperature is 25°C, the surface coefficient of heat transfer is 12 W/m²°C, and the thermal conductivity of the aluminum is 220 W/m°C. Using the concept of fin efficiency, determine the rate of heat loss from the fin per meter of pipe length.

Solution

Figures 5.11 and 5.12 present efficiencies for fins of various shapes. We first evaluate the expression

$$L_c^{3/2}\sqrt{\frac{h}{kA_m}} = \left(\frac{90+2/2}{1000}\right)^{3/2}\sqrt{\frac{12\times 10^6}{220\times 2\times(90+2/2)}} = 0.475$$

Making reference to the lower curve of Figure 5.11 we find the plate fin efficiency $\eta \approx 87\%$. Including an end correction, the surface area of the fin per meter of pipe is as follows

$$A_{\text{fin}} = \text{fin length (root to tip)} \times \text{width (top and bottom)}$$

$$= \left(\frac{90+2/2}{1000}\right)\times 2(1) = 0.182\ m^2$$

Thus, the heat loss from the fin to the surroundings will be

$$q = h(\eta_{\text{fin}} A_{\text{fin}})(T_0 - T_\infty)$$
$$= 12 \times (0.87 \times 0.182)(220 - 25) \simeq 370 \text{ W}$$

For a plate fin, we can compare the efficiency determined from Figure 5.11 with the efficiency calculated from equation 5.56.

$$m = \sqrt{\frac{hP}{kA}} = \sqrt{\frac{12 \times 2 \times 1}{220 \times 0.002 \times 1}} = 7.39$$

and

$$\eta = \frac{\tanh (mL)}{mL} = \frac{\tanh (7.39 \times 0.091)}{7.39 \times 0.091} = 0.87$$

REFERENCES

1. Gardner, K. A., *Trans. ASME, 67,* 621–631 (1945).
2. Harper, W. B., and D. R. Brown. NACA Report 158, 1922.
3. Holman, J. P., *Heat Transfer.* New York: McGraw-Hill Book Company, 1972.
4. McAdams, W. H., *Heat Transmission,* 3rd ed., p. 177. New York: McGraw-Hill Book Company, 1954.
5. Parrish, A., *Mechanical Engineer's Reference Book.* Butterworth & Co. Ltd., 1973.
6. Perry, R. H., and C. H. Chilton, *Chemical Engineer's Handbook,* 5th ed. New York: McGraw-Hill Book Company, 1973.
7. Sissom, L. E., and D. R. Pitts, *Elements of Transport Phenomena.* New York: McGraw-Hill Book Company, 1972.
8. Thomas, L. C., *Fundamentals of Heat Transfer.* Englewood Cliffs, N.J.: Prentice-Hall, Inc., 1980.

PROBLEMS

5.1. The wall of an industrial furnace is made up of 30 cm of fireclay brick, 12 cm of kaolin insulating brick, and 15 cm of masonry brick. The temperatures of the inner and outer surfaces, T_1 and T_4, are 375° and 35°C, respectively. Estimate the temperatures, T_2 and T_3, of the two intermediate surfaces. Neglect the thermal resistance of the mortar joints. *Answer:* 272°C, 113°C

5.2. A wall 20 cm thick is to be constructed of material that has an average thermal conductivity of 1.0 W/m-K. It is to be insulated with a material of average thermal conductivity 0.10 W/m-K so that the heat loss is no greater than 315 W/m². If the inner and outer surfaces of the insulated wall are at 425° and 35°C, respectively, how thick must be the layer of insulating material? *Answer:* 10 cm

5.3. A slab of material 5.0 cm thick, with a cross-sectional area of 100 cm^2, has its edges well insulated and one surface maintained at 21°C, while the other surface is kept at 122°C. A thermocouple located at the center of the slab indicates a temperature of 77°C. The steady-state heat flow through the material is 1200 W. Using these data, obtain an expression for the thermal conductivity of the material as a function of temperature. *Answer:* $k = 40.9 + 0.259\ T$ W/m-K

5.4. A composite wall made from 2.5-cm-thick steel plate, 1.5 cm of asbestos, and 1.0 cm of glass is subjected to an overall temperature difference of 110°C. Estimate the heat flux through the structure. *Answer:* 1.1×10^3 W/m^2

5.5. What is the heat flux through the composite wall shown? Assume one-dimensional heat flow. Slabs B and C have the same width, depth, and thickness.

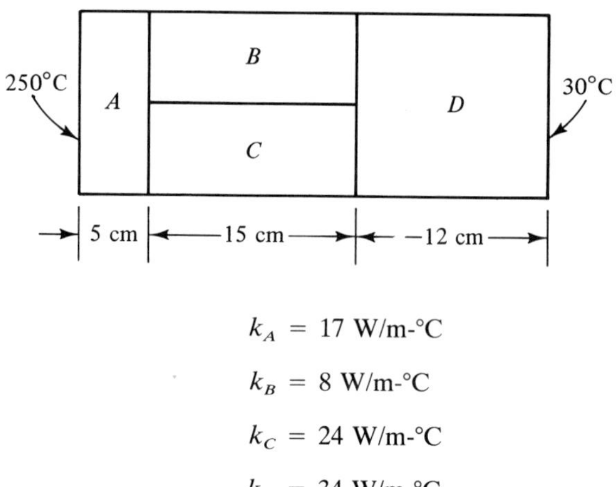

$$k_A = 17 \text{ W/m-°C}$$

$$k_B = 8 \text{ W/m-°C}$$

$$k_C = 24 \text{ W/m-°C}$$

$$k_D = 34 \text{ W/m-°C}$$

Answer: 1.4×10^4 W/m^2

5.6. A 2-in. schedule-40 steel pipe (6.033 cm OD) is covered with a 1.5-cm layer of asbestos insulation followed by a 2.0-cm layer of fiber glass insulation. The temperature of the pipe wall is 270°C and 25°C for the outside surface of the insulation. Estimate the temperature at the interface between the asbestos and fiber glass. *Answer:* 191°C

5.7. A 5-in. schedule-80 steel pipe (14.130 cm OD, wall thickness 0.953 cm) is covered by a 3.0-cm layer of high-temperature insulation (k = 0.087 W/m-°C), followed by a 3.0-cm layer of 85% magnesia. The temperature of the inside surface of the steel pipe is 340°C, and for the outside surface of the magnesia it is 22°C.

(a) What is the rate of heat loss per unit length from the insulated pipe? *Answer:* 245 W/m^2

(b) If the two insulating materials were interchanged what would be the heat loss from the pipe? *Answer:* 234 W/m^2

5.8. A $\frac{3}{8}$-in. copper hot-water line (1.270 cm OD) is to be insulated with a 1.30-cm layer of material that has a thermal conductivity of 0.15 W/m-°C. The temperature of the copper wall may be taken as that of the water, which is 90°C. The air surrounding the line has an ambient temperature of 21°C, and the air convection coefficient is estimated to be 5.7 W/m^2-°C. By what percentage has the heat loss from the line been reduced by the layer of insulation? *Answer:* -68%

5.9. A steel pipe (k = 62 W/m-°C) carrying cold brine has an inside diameter of 5.08 cm and an outside diameter of 6.10 cm. The pipe is covered with 1.30 cm of lead (k = 35 W/m-°C). Moist air with a dew point of 4.4°C passes over the pipe and forms a layer of ice (k = 2.3 W/m-°C) on the lead. Before any ice has formed, the temperature of the lead surface is determined to be −25°C. The temperature of the inner surface of the pipe remains constant at −28°C. When the layer of ice has reached maximum thickness, the heat transfer to the pipe is found to be 90% of its value when no ice had formed. Estimate the thickness of the layer of ice. *Answer:* 1.42 cm

5.10. A vertical stack 80 ft high has an inside diameter of 18 in. It is constructed with an inside layer of heat-resistant brick (k = 1.5 Btu/hr-ft-F°) that is 12 in. thick, an adjacent layer of building brick (k = 0.5 Btu/hr-ft-F°) that is also 12 in. thick, and an outer steel shell (k = 36 Btu/hr-ft-F°) that is 1/2 in. thick. Air at 1 atm pressure and 1000°F enters the base of the stack at a velocity of 10 ft/sec. If the average temperature of the inside surface of the stack is 700°F and for the outside surface of the steel shell it is 80°F, at what temperature will the air leave the stack? *Answer:* 490°F

5.11. A 2-in. schedule-40 steel pipe (2.067 in. ID, 2.375 in. OD, k = 30 Btu/hr-ft-F°) carries steam at 300°F. It is lagged with 1/2 in. of rock wool (k = 0.033 Btu/hr-ft-F°). The surrounding ambient air temperature is 70°F, and the convection coefficient is calculated to be 2.23 Btu/hr-ft^2-F°). Estimate the heat loss per foot length of pipe. *Answer:* 104 Btu/hr-ft

5.12. A spherical tank containing liquid nitrogen at 1 atm pressure is insulated with a material having a very low thermal conductivity ($k = 1.73 \times 10^{-3}$ W/m-°C). The inside diameter of the tank is 60 cm, and the insulation thickness is 2.5 cm. Estimate the kilograms of nitrogen vaporized per day if the outside surface of the insulation is at 21°C. The normal boiling point of nitrogen is −196°C, and the latent heat of vaporization of nitrogen is 200 kJ/kg. *Answer:* 8.0 kg/day

5.13. A plane wall 10 cm thick with a thermal conductivity of 26 W/m-°C is well insulated on one side and in contact with the environment at an ambient temperature of 40°C on the other side. The convection coefficient at the exposed surface is estimated to be 280 W/m^2-°C. If heat is generated within the wall at a rate of 50,000 W/m^3, estimate the maximum value of the temperature in the wall. *Answer:* 68°C

5.14. A 1.60-mm diameter stainless steel rod that is 60 cm long has a potential of 15 V placed across the ends. The rod is immersed in water boiling at 100°C. The convection coefficient of heat transfer is estimated to be 6800 W/m^2-°C. Estimate the maximum value of the temperature in the wire. Take the resistivity of stainless steel as 70 μΩ-cm and the thermal conductivity as 22 W/m-°C. *Answer:* 159°C

5.15. For the plane wall in Problem 5.13, suppose all conditions remain as given except that both surfaces are uninsulated and exposed to the environment at

40°C with a convection coefficient of 280 W/m²-°C. What is the maximum value of the temperature in the wall? *Answer:* 51.3°C

5.16. If conditions remain as specified in Problem 5.15 except that the ambient temperature to which *one* surface is exposed is increased to 50°C, determine the two surface temperatures, the maximum value of the temperature, and its location within the wall. *Answer:* $T_{w1} = 52.2$°C, $T_{w2} = 55.7$°C, $T_0 = 56.65$°C, $x_0 = 1.82$ cm

5.17. A wall at 230°C has an aluminim rod 20 mm in diameter and 15 cm long protruding from it. The surrounding air is at 21°C, and the convection coefficient at the rod surface is 14 W/m²-°C. Estimate the rate at which the rod transfers heat to the air by convection. *Answer:* 26 W

5.18. Two walls, one at 230°C and the other at 150°C, are connected by a 15-mm-diameter steel rod that is 30 cm long. Air at 21°C surrounds the rod, and the convection coefficient at the rod surface is 15 W/m²-°C. Estimate the total rate of heat transfer by convection from the rod to the air. *Answer:* 22.5 W

5.19. An angular copper fin, 1.60 mm wide by 6.50 mm long, is bonded to a tube of the same metal, which is 2.50-cm OD. The temperature of the tube wall is 125°C, and for the surrounding air it is 21°C; the convection coefficient is 17 W/m²-°C. Estimate the rate at which heat is transferred to the air by the fin. *Answer:* 2.62 W

5.20. A 2.5-cm-OD tube has annular fins of rectangular profile spaced at 1-cm increments along its length. The fins are of aluminum and are 0.80 mm thick and 13 mm long. For a tube wall temperature of 150°C, a surrounding air temperature of 22°C, and a convection coefficient of 115 W/m²-°C, estimate the rate of heat transfer from the tube per meter of length. *Answer:* 5200 W/m

5.21. Two long rods of copper, 15 mm in diameter, are soldered end to end. The melting point of the solder is 650°C. Determine the minimum energy input to the soldered joint required to maintain the temperature at 650°C. The ambient air temperature is 21°C and the convection coefficient between rods and the air is 14 W/m²-°C. *Answer:* 260 W

5.22. A heat exchanger for cooling a small engine is to use rectangular fins made of an aluminum alloy (k = 216 W/m-°C). The fins are to be 4.0 cm long and (a) either 0.81 mm thick spaced 6.4 mm apart, or (b) 1.62 mm thick spaced 12.8 mm apart. The exchanger wall temperature is 121°C, and the ambient air temperature may be taken as 21°C. Which design will result in the highest rate of heat transfer per unit exchanger area? The convection coefficient may be taken as 14 W/m²-°C. *Answer:* 15,000 W/m², design (a)

Convective Heat Transfer

6

6.1 Introduction

In Chapter 5 we introduced the subject of one-dimensional steady-state heat conduction. The objective was the presentation and use of Fourier's heat conduction equation.

$$q = -kA \frac{dT}{dx} \tag{6.1}$$

However, to consider realistic problems, we could not for long ignore the presence of heat transfer at fluid–solid interfaces. Heat conducted through a solid must eventually find its way to the surroundings and, except in the case of internal generation, the heat supply had to come from the surroundings. This led to the early introduction of the concept of a surface coefficient of heat transfer and calculation of heat transfer by the following equation, known as Newton's law of cooling.

$$q = hA(T_w - T_\infty) \tag{6.2}$$

Evaluation of the surface coefficient of heat transfer h now becomes our focus of attention.

A hot object immersed in a fluid will raise the temperature of the adjacent fluid by conduction heat transfer. The buoyant force caused by the change of density and the influence of gravity carries the heated fluid away, and it is replaced by fresh fluid. This process is called *free* or *natural convective heat transfer*. Everyday experience tells us that the rate of convective heat transfer can be increased by immersing the heated object in a fast-moving fluid stream. Heat transfer from a solid surface into an already flowing fluid is called *forced convective heat transfer*. In either situation, free or forced convection, the most direct method to evaluate the surface coefficient h is by experiment. In some cases, h can be found by mathematical analysis of events in the fluid boundary layer.

6.2 First Law Method for Finding *h*

A measured rate of energy input can be used to keep a solid at a steady elevated temperature. An equal amount of energy must escape from the solid surface into the surrounding fluid. If the surface area is known, and the surface temperature and fluid bulk temperature at a distance are measured, the information can be used with Newton's law of cooling to determine h.

$$h = \frac{q}{A(T_w - T_\infty)} \tag{6.3}$$

An alternative method to find h is to preheat the solid, immerse it in fluid surroundings, and measure the rate of change of temperature of the solid while it cools. The method is illustrated by the following example.

6.2.1 Cooling a Metal Sphere. Consider a metal sphere, preheated to temperature T_0, and suspended in still air. Through the process of natural convective heat transfer, energy flows from the sphere to the surrounding air, and the temperature of the sphere drops (Figure 6.1).

We can assume uniform temperature throughout the sphere provided the metal has a high thermal conductivity and the rate of heat loss from the surface is not too rapid. At any specified time, the rate of heat transfer from

Figure 6.1 Temperature–time history for a sphere cooling

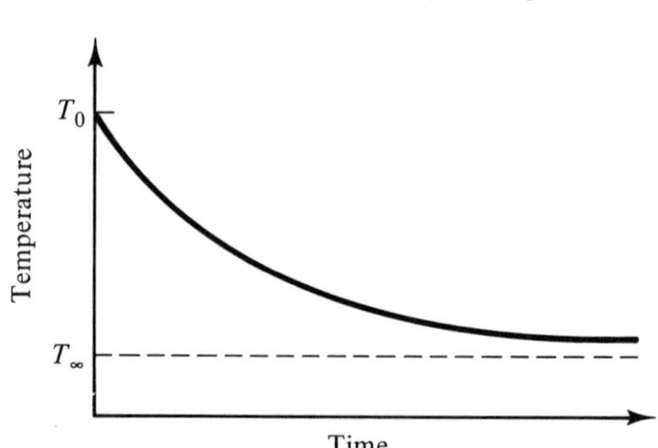

the surface of the sphere must equal the rate of decrease of stored energy in the sphere.

$$q = -mc_p \frac{dT_w}{dt} = hA(T_w - T_\infty) \tag{6.4}$$

where m = mass of sphere

c_p = specific heat of metal

A = surface area of sphere

T_w = surface temperature of sphere

T_∞ = temperature of air at a distance

The instantaneous value of the surface coefficient when the sphere is at temperature T is given by

$$h = \frac{-mc_p}{A(T_w - T_\infty)} \frac{dT_w}{dt} \tag{6.5}$$

Values of T_w and dT_w/dt can be read directly from the temperature–time curve (Figure 6.1) and used in equation (6.5) to determine h.

6.3 Temperature Gradient at Solid–Fluid Interface

As is characteristic of many thermodynamic analyses, the first law experimental method required no detailed knowledge of the surface heat-transfer mechanism because the heat flux was known, independent of the heat-transfer mechanism. In the following method, the heat flux and surface coefficient h are determined through knowledge of the rate process occurring at the solid–fluid interface. Fluid in immediate contact with a solid surface adheres to that surface because of viscosity and is therefore at rest. Thus, heat flux from the solid must enter the fluid by pure conduction. From this argument we can formulate the following equation:

$$\frac{q}{A} = h(T_w - T_\infty) = -k\left(\frac{dT}{dy}\right)_{\text{at } y=0} \tag{6.6}$$

where k = thermal conductivity of the fluid

y = distance measured from the solid surface into the fluid

Hence

$$h = -\frac{k}{(T_w - T_\infty)} \left(\frac{dT}{dy}\right)_{\text{at } y=0} \tag{6.7}$$

By equation (6.7), we are stating that the problem is solved if we know the slope of the temperature gradient in the fluid at the solid surface, the thermal conductivity of the fluid, and the overall temperature difference $T_w - T_\infty$.

In principle, we might measure the fluid temperature with a temperature-sensing probe supported on a micrometer drive so that the variation of fluid temperature with distance near the wall is determined. In practice, this temperature measurement is made difficult by radiation effects, heat conduction along the probe, and so on.

The optical interferometer is a laboratory apparatus used to determine fluid temperatures in the vicinity of a heated object. A light beam is split into two, with one of the beams directed past the heated surface where it is altered by the changing index of refraction of the fluid, which in turn is dependent on the fluid temperature. When the split beams are brought back together and projected onto a viewing screen, a pattern of interference fringes is displayed, such as shown in Figure 6.2. The spacing of the fringes, which are lines of constant temperature, can be measured at any location on the surface, and the local temperature profile can be displayed, as in Figure 6.3.

In Figure 6.3, the actual thermal boundary layer thickness δ_T is shown, based on the measured temperature profile. A pseudothickness δ_T' can be defined assuming a linear temperature profile with slope equal to $(dT/dy)_{y=0}$.

Figure 6.2 Interference fringes in air surrounding a heated metal block

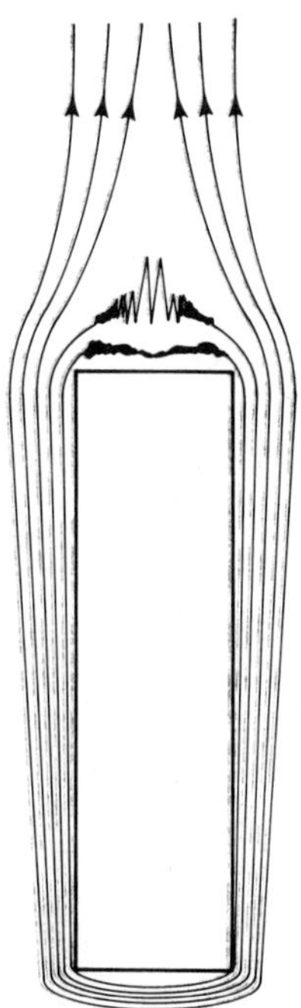

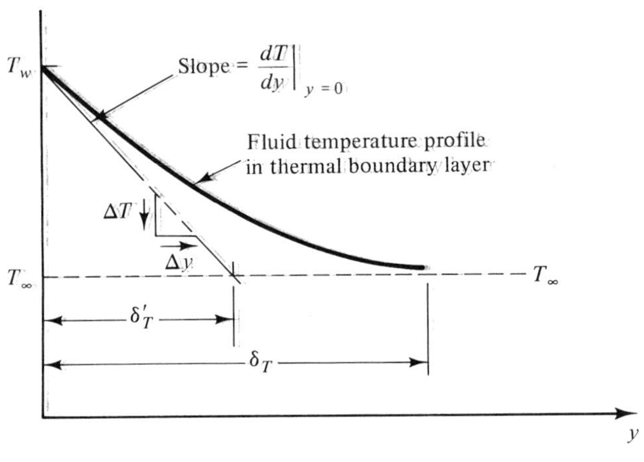

Figure 6.3 Thermal boundary layer temperature profile

From similar triangles, we note that

$$\left.\frac{dT}{dy}\right|_{y=0} = \frac{\Delta T}{\Delta y} = \frac{T_\infty - T_w}{\delta_T'} \tag{6.8}$$

Inserting this expression into equation (6.7), we get, for the surface coefficient

$$h = -\frac{k}{T_w - T_\infty}\frac{T_\infty - T_w}{\delta_T'} = \frac{k}{\delta_T'} \tag{6.9}$$

For a specified temperature profile shape, the pseudothickness δ_T' is a fixed fraction of the temperature boundary layer thickness δ_T. We can thus interpret equation (6.9) to mean that h is proportional to the fluid thermal conductivity and inversely proportional to the thickness of the temperature boundary layer.

6.4 Determination of *h* Using Boundary Layer Analysis

In Sections 6.2 and 6.3, we saw some experimental methods for finding the surface coefficient of heat transfer. In addition, we were introduced to integral methods of boundary layer analysis in Chapter 4. We propose now to continue the boundary layer analysis to determine h, as an alternative to the experimental methods already discussed.

6.4.1 Heat Flux Equation. If we make the assumption that fluid properties like density and viscosity are unaffected by temperature, we can determine the boundary layer thickness and drag without reference to temper-

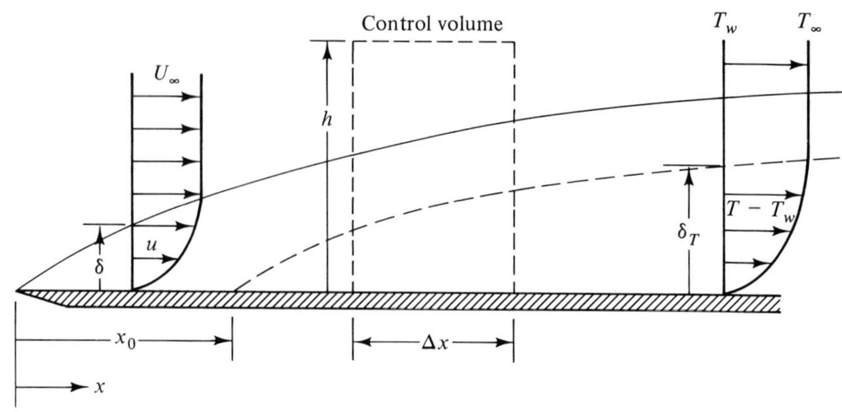

Figure 6.4 Velocity and temperature boundary layers

ature and heat transfer. Then, with the velocity boundary layer known, we can consider the temperature boundary layer and heat transfer. Let the control volume height h be greater than δ or δ_T, and the control volume width b be normal to Figure 6.4. Consider the energy quantities crossing the control volume boundary. Assume kinetic energies are relatively small.

$$\text{Mass flow entering left side} = b \int_0^h \rho u \, dy$$

$$\text{Enthalpy entering left side} = b \int_0^h \rho u c_p T \, dy$$

$$\text{Mass flow leaving right side} = b \int_0^h \rho u \, dy + \frac{d}{dx}\left(b \int_0^h \rho u \, dy\right)\Delta x$$

$$\text{Enthalpy leaving right side} = b \int_0^h \rho u c_p T \, dy + \frac{d}{dx}\left(b \int_0^h \rho u c_p T \, dy\right)\Delta x$$

Mass flow entering top of control volume

= mass flow leaving right side − mass flow entering left side

$$= \frac{d}{dx}\left(b \int_0^h \rho u \, dy\right)\Delta x$$

$$\text{Enthalpy entering top} = c_p T_\infty \frac{d}{dx}\left(b \int_0^h \rho u \, dy\right)\Delta x$$

$$\text{Heat transfer entering bottom} = -kb \, \Delta x \left.\frac{\partial T}{\partial y}\right|_{y=0}$$

For steady flow, the principle of energy conservation requires that the energy flow rate into the left side, top, and bottom be equal to the energy flow out the right side.

$$b \int_0^h \rho u c_p T \, dy + c_p T_\infty \frac{d}{dx}\left(b \int_0^h \rho u \, dy\right)\Delta x - kb\,\Delta x \left.\frac{\partial T}{\partial y}\right|_{y=0}$$

$$= b \int_0^h \rho u c_p T \, dy + \frac{d}{dx}\left(b \int_0^h \rho u \, c_p T \, dy\right)\Delta x \qquad (6.10)$$

Canceling like terms and rearranging equation (6.10) leads to the *heat flux equation*,

$$\frac{d}{dx}\left[\int_0^h (T_\infty - T)u \, dy\right] = \frac{k}{\rho c_P} \left.\frac{\partial T}{\partial y}\right|_{y=0} \qquad (6.11)$$

If the velocity boundary layer has already been solved, independent of heat transfer, we can now use the heat flux equation to solve for the temperature boundary layer and rate of heat transfer.

6.4.2 Heat Transfer from a Flat Plate. Consider Figure 6.4 and assume that the velocity boundary layer and the temperature boundary layer shapes will each be represented by third-degree polynomials. Heat transfer begins at distance x_0 downstream from the leading edge of the plate. Using the momentum integral equation, we will determine the rate of growth of the velocity boundary layer. Then the heat flux equation will be used to calculate the thickness of the temperature boundary layer at any x. From the temperature gradient at the wall, we can then calculate the heat transfer.

For flow over a flat plate with no pressure gradient, the momentum integral is given by equation (4.24).

$$\frac{d\delta_2}{dx} = \frac{\tau_w}{\rho U_\infty^2} \qquad (4.24)$$

From equation (4.36), the velocity profile is described by

$$\frac{u}{U_\infty} = \frac{3}{2}\eta - \frac{1}{2}\eta^3 \qquad (4.36)$$

Now

$$\tau_w = \mu \left.\frac{\partial u}{\partial y}\right|_{y=0} = \frac{\mu U_\infty}{\delta} \left.\frac{\partial (u/U_\infty)}{\partial \eta}\right|_{\eta=0} = \frac{3}{2}\frac{\mu U_\infty}{\delta}$$

The defining equation for momentum thickness (4.2) gives the ratio δ_2/δ for the chosen profile shape.

$$\frac{\delta_2}{\delta} = \int_{\eta=0}^{1} \frac{u}{U_\infty}\left(1 - \frac{u}{U_\infty}\right) d\eta \qquad (4.2)$$

$$= \int_0^1 \left(\frac{3}{2}\eta - \frac{1}{2}\eta^3\right)\left(1 - \frac{3}{2}\eta + \frac{1}{2}\eta^3\right) d\eta$$

$$= \frac{39}{280}$$

Substituting the values for τ_w and δ_2 into the momentum integral equation, we have

$$\frac{d}{dx}\left(\frac{39}{280}\delta\right) = \frac{1}{\rho U_\infty^2}\frac{3}{2}\frac{\mu U_\infty}{\delta} = \frac{3}{2}\frac{\mu}{\rho U_\infty \delta}$$

$$\delta\, d\delta = \frac{280}{39} \times \frac{3}{2}\frac{\mu}{\rho U_\infty}\, dx = \frac{140}{13}\frac{\mu}{\rho U_\infty}\, dx \tag{6.12}$$

$$\delta^2 = \frac{280}{13}\frac{\mu}{\rho U_\infty} x + c \tag{6.13}$$

(Note that $\delta = 0$ at $x = 0$ and therefore $c = 0$.) The temperature boundary layer profile is assumed to be

$$\frac{T - T_w}{T_\infty - T_w} = \frac{3}{2}\frac{y}{\delta_T} - \frac{1}{2}\left(\frac{y}{\delta_T}\right)^3 \tag{6.14}$$

The following quantities in the heat flux equation are derived from the assumed temperature profile.

$$T_\infty - T = T_\infty - T_w + T_w - T$$

$$= (T_\infty - T_w)\left(1 - \frac{T - T_w}{T_\infty - T_w}\right)$$

$$= (T_\infty - T_w)\left[1 - \frac{3}{2}\frac{y}{\delta_T} + \frac{1}{2}\left(\frac{y}{\delta_T}\right)^3\right] \tag{6.15}$$

$$\left.\frac{\partial T}{\partial y}\right|_{y=0} = (T_\infty - T_w)\left(\frac{3}{2\delta_T} - \frac{3}{2}\frac{y^2}{\delta_T^3}\right)_{y=0} = \frac{3}{2}\frac{T_\infty - T_w}{\delta_T} \tag{6.16}$$

Inserting these quantities into the heat flux equation (6.11),

$$\frac{d}{dx}\left\{\int_0^h (T_\infty - T_w)\left[1 - \frac{3}{2}\frac{y}{\delta_T} + \frac{1}{2}\left(\frac{y}{\delta_T}\right)^3\right] U_\infty \left[\frac{3y}{2\delta} - \frac{1}{2}\left(\frac{y}{\delta}\right)^3\right] dy\right\} = \frac{3}{2}\frac{k}{\rho c_p}\frac{T_\infty - T_w}{\delta_T} \tag{6.17}$$

The upper integration limit can be changed from h to δ_T since the quantity $T_\infty - T$ is zero in the range from δ_T to h. Canceling the term $T_\infty - T_w$ from each side, carrying out the integration between 0 and δ_T, substituting $\Delta = \delta_T/\delta$, and noting that U_∞ is a constant, the heat flux equation produces the following result:

$$U_\infty \frac{d}{dx}\left(\frac{3}{20}\delta\Delta^2 + \frac{3}{280}\delta\Delta^4\right) = \frac{3}{2}\frac{k}{\rho c_p \Delta\delta} \tag{6.18}$$

Equation (6.18) is further simplified by neglecting the term $(3/280)\,\delta\Delta^4$ (based on the small coefficient 3/280 and the assumption that Δ is less than 1).

Expanding the remaining derivative and multiplying by $\Delta\delta$,

$$\frac{3}{20} U_\infty \Delta\delta \left(2\delta\Delta \frac{d\Delta}{dx} + \Delta^2 \frac{d\delta}{dx} \right) = \frac{3}{2} \frac{k}{\rho c_p} \tag{6.19}$$

Inserting values from equations (6.12) and (6.13) for $\delta \, d\delta$ and δ^2,

$$\frac{U_\infty}{10} \left(\frac{2 \times 280}{13} \frac{\mu x}{\rho U_\infty} \Delta^2 \frac{d\Delta}{dx} + \frac{140}{13} \Delta^3 \frac{\mu}{\rho U_\infty} \right) = \frac{k}{\rho c_p} \tag{6.20}$$

$$\frac{14}{13} \frac{\mu x}{\rho} \left(4\Delta^2 \frac{d\Delta}{dx} + \frac{\Delta^3}{x} \right) = \frac{k}{\rho c_p}$$

$$\frac{14}{13} \frac{\mu x}{\rho} \left(\frac{4}{3} \frac{d\Delta^3}{dx} + \frac{\Delta^3}{x} \right) = \frac{k}{\rho c_p}$$

Arranging this first-order differential equation in standard form,

$$\frac{d\Delta^3}{dx} + \frac{3}{4x} \Delta^3 = \frac{3}{4} \frac{13}{14} \frac{k}{\mu c_p x} \tag{6.21}$$

Multiplying by the integrating factor $x^{3/4}$,

$$\frac{d}{dx} (x^{3/4} \, \Delta^3) = \frac{3}{4} \frac{13}{14} \frac{k}{\mu c_p} x^{-1/4} \tag{6.22}$$

[*Note:* The student might check that this derivative is equivalent to the two terms in (6.21) $\times$ integrating factor.]

Integrating equation (6.22),

$$x^{3/4} \, \Delta^3 = \frac{3}{4} \times \frac{13}{14} \frac{k}{\mu c_p} \times \frac{4}{3} x^{3/4} + c$$

$$= \frac{13}{14} \frac{k x^{3/4}}{\mu c_p} + c$$

$$\Delta^3 = \frac{13}{14} \frac{k}{\mu c_p} + \frac{c}{x^{3/4}}$$

For the condition that $\Delta = 0$ at $x = x_0$

$$c = -\frac{13}{14} \frac{k}{\mu c_p} x_0^{3/4}$$

Thus

$$\Delta^3 = \frac{13}{14} \frac{k}{\mu c_p} \left[1 - \left(\frac{x_0}{x} \right)^{3/4} \right]$$

$$\frac{\delta_T}{\delta} = \left\{ \frac{13}{14} \frac{k}{\mu c_p} \left[1 - \left(\frac{x_0}{x} \right)^{3/4} \right] \right\}^{1/3} \tag{6.23}$$

If the whole plate is heated, $x_0 = 0$:

$$\frac{\delta_T}{\delta} = \left(\frac{13}{14}\frac{k}{\mu c_p}\right)^{1/3} = \frac{\mathrm{Pr}^{-1/3}}{1.025} \tag{6.24}$$

From equation (6.13), we have

$$\delta = \sqrt{\frac{280}{13}\frac{\mu x}{\rho U_\infty}} = 4.64x\ \mathrm{Re}_x^{-1/2} \tag{6.25}$$

Multiplying equation (6.24) by (6.25),

$$\delta_T = 4.53x\ \mathrm{Pr}^{-1/3}\ \mathrm{Re}_x^{-1/2} \tag{6.26}$$

From equation (6.7),

$$h_{\mathrm{local}} = -\frac{k(dT/dy)_{y=0}}{T_w - T_\infty} \tag{6.7}$$

and from the assumed temperature profile (6.14),

$$\frac{(dT/dy)_{y=0}}{T_\infty - T_w} = \frac{3}{2\delta_T}$$

Thus

$$h_{\mathrm{local}} = \frac{3}{2}\frac{k}{\delta_T} \tag{6.27}$$

Substituting equation (6.26) into (6.27),

$$h_{\mathrm{local}} = 0.331\frac{k}{x}\ \mathrm{Pr}^{1/3}\ \mathrm{Re}_x^{1/2} \tag{6.28}$$

Arranged in nondimensional form,

$$\frac{h_{\mathrm{local}}x}{k} = 0.331\mathrm{Pr}^{1/3}\ \mathrm{Re}_x^{1/2} \tag{6.29}$$

The group hx/k is called the *Nusselt number*, Nu_x. Equation (6.28) gives the local surface coefficient at location x along the flat plate. To determine the average value of h for the whole plate from $x = 0$ to $x = 1$, we set up a differential expression for heat transfer at location x and integrate over the length of the plate. Let δq represent the heat flow rate from a plate strip b units wide and dx long.

$$\delta q = h \times \text{Area} \times \Delta T = h_{\mathrm{local}} \times b \times dx(T_w - T_\infty) \tag{6.30}$$

For the whole plate,

$$q = \int_{x=0}^{x=l} h_{\mathrm{local}} \times b \times dx(T_w - T_\infty)$$

$$= b(T_w - T_\infty)\int_0^l h_{\mathrm{local}}\,dx \tag{6.31}$$

By Newton's law of cooling, we also have for the whole plate

$$q = h_{\text{ave}} \times bl(T_w - T_\infty) \tag{6.32}$$

Equating equations (6.31) and (6.32),

$$h_{\text{ave}} = \frac{1}{l}\int_0^l h_{\text{local}}\, dx \tag{6.33}$$

Use (6.28) for h_{local}, noting that $\sqrt{\text{Re}_x}/x = \sqrt{\rho U_\infty/\mu x}$,

$$\begin{aligned} h_{\text{ave}} &= \frac{1}{l}\int_0^l 0.331k\ \text{Pr}^{1/3}\sqrt{\frac{\rho U_\infty}{\mu}} \times \frac{dx}{\sqrt{x}} \\ &= \frac{1}{l}\, 0.331k\ \text{Pr}^{1/3}\sqrt{\frac{\rho U_\infty}{\mu}}\, 2\sqrt{x}\,\Big|_0^l \\ &= 0.662\,\frac{k}{l}\ \text{Pr}^{1/3}\ \text{Re}_l^{1/2} \end{aligned} \tag{6.34}$$

The Nusselt number for the plate is

$$\text{Nu} = \frac{h_{\text{ave}} l}{k} = 0.662\text{Pr}^{1/3}\ \text{Re}_l^{1/2} \tag{6.35}$$

The average surface coefficient for the whole plate is twice the local value at $x = l$.

6.5 Suggested Perspective

After working through the intricacies of the integral solution for heat transfer from a flat plate (Section 6.4.2), it may be appropriate to draw the student's attention back to the main theme of this chapter. In fact, we urge the student to remain aware of the overall objective, even if some of the detail is obscure.

Summarizing, we have noted that the solution of realistic heat transfer problems will involve convective heat transfer, either natural or forced. Superficially, we might consider that Newton's law of cooling ($q = hA\ \Delta T$) solves the convective aspect of the heat-transfer problem, but in fact it substitutes a new problem, which is the selection of a suitable value for h. The researcher in heat transfer has several methods for evaluating h for particular configurations, and Sections 6.2 through 6.4 should provide some idea of how this may be done.

On the other hand, the engineer who designs or estimates performance of heat-transfer apparatus will not in general be using these methods. Values of h will be used that have been determined and compiled by others. We will now look at how the data get organized for presentation and end use. Finally, we will present a number of correlations from which h can be found.

6.6 Forced Convection Correlation

Once again we encounter a situation where the desired quantity, h in this case, is a function of many variables. Dimensional analysis is used to reduce the number of independent variables and to provide hints about the form that correlations should take. For a more complete discussion of the subject than will be presented here, the student is advised to review Section 3.4.

We can expect the average surface coefficient h for an object of known geometry to be affected by the following variables: fluid density, velocity, viscosity, specific heat, and thermal conductivity, and a characteristic length for the object.

$$h = f(\rho, V, \mu, c_p, k, l) \tag{6.36}$$

Assume this functional relation can be expressed as a power series:

$$\begin{aligned} h = {} & c_1(\rho^{a1} V^{b1} \mu^{c1} c_p{}^{d1} k^{e1} l^{f1}) \\ & + c_2(\rho^{a2} V^{b2} \mu^{c2} c_p{}^{d2} k^{e2} l^{f2}) \\ & + c_3 \cdots \end{aligned} \tag{6.37}$$

Each term in the series must have the same dimensions as h. Taking a typical term and dropping the subscript on the exponents for convenience,

$$h \overset{D}{=} \rho^a V^b \mu^c c_p{}^d k^e l^f \tag{6.38}$$

Using mass, length, time, and temperature as our basic dimensions, the dimensions and units (SI) of the various quantities are given in Table 6.1.

We can balance dimensions on either side of equation (6.38).

	Left side	$\overset{D}{=}$	Right side	
M	1	=	$a + c + e$	(6.39)
L	0	=	$-3a + b - c + 2d + e + f$	(6.40)
t	-3	=	$-b - c - 2d - 3e$	(6.41)
T	-1	=	$-\mathrm{d} - \mathrm{e}$	(6.42)

TABLE 6.1

Quantity	Units	Dimensions
h	W/m²- K	$Mt^{-3}T^{-1})$
ρ	kg/m³	ML^{-3}
V	m/s	Lt^{-1}
μ	kg/m-s	$ML^{-1}t^{-1}$
c_p	J/kg-K	$L^2t^{-2}T^{-1}$
k	W/m-K	$MLt^{-3}T^{-1}$
l	m	L

With six unknowns and four equations, we can reduce the number of unknowns to two. Solving in terms of a and e, we obtain

$$d = 1 - e$$

$$c = 1 - a - e$$

$$b = a$$

$$f = a - 1$$

Substitute into equation (6.38):

$$\begin{aligned} h &\stackrel{D}{=} \rho^a V^a \mu^{1-a-e} c_p{}^{1-e} k^e l^{a-1} \\ &\stackrel{D}{=} \left(\frac{\mu c_p}{l}\right)^1 \left(\frac{\rho V l}{\mu}\right)^a \left(\frac{k}{\mu c_p}\right)^e \end{aligned} \tag{6.43}$$

We therefore have a series of terms in equation (6.37) as follows:

$$\begin{aligned} h &= c_1\left(\frac{\mu c_p}{l}\right)\left(\frac{\rho V l}{\mu}\right)^{a1}\left(\frac{k}{\mu c_p}\right)^{e1} + c_2\left(\frac{\mu c_p}{l}\right)\left(\frac{\rho V l}{\mu}\right)^{a2}\left(\frac{k}{\mu c_p}\right)^{e2} + \cdots \\ &= \frac{\mu c_p}{l}\left[c_1\left(\frac{\rho V l}{\mu}\right)^{a1}\left(\frac{k}{\mu c_p}\right)^{e1} + c_2\left(\frac{\rho V l}{\mu}\right)^{a2}\left(\frac{k}{\mu c_p}\right)^{e2} + \cdots\right) \\ &= \frac{\mu c_p}{l} f\left(\frac{\rho V l}{\mu}, \frac{k}{\mu c_p}\right) \end{aligned}$$

or

$$\frac{hl}{\mu c_p} = f\left(\frac{\rho V l}{\mu}, \frac{k}{\mu c_p}\right) \tag{6.44}$$

So that relation (6.44) is expressed in terms of recognized nondimensional numbers, we divide the left side by the group $k/\mu c_p$, and we invert this group on the right side.

$$\frac{hl}{k} = f\left(\frac{\rho V l}{\mu}, \frac{\mu c_p}{k}\right) \tag{6.45}$$

We have shown that for forced convective heat transfer the Nusselt number hl/k will be a function of Reynolds number, $\rho V l/\mu$, and the Prandtl number for the fluid, $\mu c_p/k$.

$$\text{Nu} = f(\text{Re}, \text{Pr}) \tag{6.46}$$

This conclusion is in agreement with equation (6.35), which expresses the Nusselt number for a flat plate with forced convection.

6.7 Natural Convection Correlation

Recollect that under natural convection conditions there is no externally imposed velocity. Any fluid movement results from the buoyancy of the heated field. Thus in our selection of variables to be considered we must replace the velocity variable, which we had in the case of forced convection, with a quantity representing the buoyancy effect. The rate of change of fluid density with temperature is usually expressed in terms of a volume expansion coefficient β. By definition,

$$\beta = \frac{1}{v}(\partial v/\partial T)_p = -\frac{1}{\rho}(\partial \rho/\partial T)_p$$

The magnitude of the buoyancy effect is proportional to the gravitational field g, the volume expansion coefficient β, and the temperature difference $T - T_\infty$.

Thus, in the list of significant variables for natural convection, we will use the same list as for forced convection and replace V with the quantity ($g\beta$ ΔT). The dimensional equation for one sample term of the power series is

$$H \overset{D}{=} \rho^a(g\beta\,\Delta T)^b \mu^c c_p{}^d k^e l^f \tag{6.47}$$

Note that the product $\beta\,\Delta T$ is dimensionless, so the quantity $g\beta\,\Delta T$ has dimensions Lt^{-2}. We can proceed to balance dimensions M, L, t, and T as in Section 6.6, or we might simply observe that $\sqrt{lg\beta\,\Delta T}$ has dimensions Lt^{-1} the same as velocity. In either case, the result is

$$h = \frac{\mu c_p}{l} \times f\left(\frac{\rho l\sqrt{lg\beta\,\Delta T}}{\mu}, \frac{k}{\mu c_p}\right) \tag{6.48}$$

Rearranging equation (6.48) in the standard form,

$$\frac{hl}{k} = f\left(\frac{g\beta l^3\,\Delta T}{\nu^2}, \frac{\mu c_p}{k}\right) \tag{6.49}$$

For natural convection, the Nusselt number hl/k is a function of the Grashof number, $g\beta l^3\,\Delta T/\nu^2$, and the Prandtl number for the fluid, $\mu c_p/k$.

$$\text{Nu} = f(\text{Gr}, \text{Pr}) \tag{6.50}$$

6.8 Empirical Correlations for Convective Heat Transfer

Numerous correlations have been developed for a variety of common configurations, based on the experimental results of many researchers in heat transfer. In specialized texts on the subject, limiting conditions under which each correlation may be used are carefully spelled out, and often the range of possible error is also given. With this full information, the discriminating user can carefully select the best correlation for a particular problem.

Our objective in this section is more limited. For the beginner, there is sufficient challenge in extracting the numerical value of h from a given correlation and using this value to solve the given problem. The correlations to be presented are just a small sampling of those available.

The fluid properties used in the correlations, such as density, viscosity, and thermal conductivity, are generally temperature dependent, and of course the temperature itself varies in the field from the temperature at the wall to the temperature of the fluid at a distance. Two commonly used reference temperatures are the fluid bulk temperature and the fluid film temperature. For flow in pipes, the bulk temperature at a particular location in the pipe is the average temperature if the fluid was thoroughly mixed, sometimes called the cup-mixing temperature. The film temperature for flow in pipes is the average of the wall temperature and the bulk temperature. The arithmetic mean bulk temperature over a length of pipe is the average of the entering and leaving bulk temperatures. For flow over a plate or around an object, the film temperature is the average of the wall temperature and T_∞. The concept of bulk temperature is not useful in this case.

6.8.1 Some Correlations for Forced Convection

Laminar Flow in Pipes E. N. Sieder and G. E. Tate (1) developed the following correlation for the average surface coefficient, with fluid properties except μ_w evaluated at the arithmetic mean bulk temperature.

$$\mathrm{Nu} = 1.86\mathrm{Re}^{1/3}\,\mathrm{Pr}^{1/3}\left(\frac{D}{L}\right)^{1/3}\left(\frac{\mu}{\mu_w}\right)^{0.14} \tag{6.51}$$

The viscosity μ_w is evaluated at the wall temperature. For this correlation, h is based on the average of the inlet and outlet values of $T_{\text{bulk}} - T_w$.

Fully Developed Turbulent Flow in Pipes For correlations (6.52) through (6.55) inclusive, fluid properties are evaluated at the arithmetic mean bulk temperature.

The correlation due to F. W. Dittus and L. M. K. Boelter (1) is commonly used for flow in smooth-walled tubes with moderate temperature differences.

$$\mathrm{Nu} = 0.023\mathrm{Re}^{0.8}\,\mathrm{Pr}^{m} \tag{6.52}$$

where $m = 0.4$ for heating

$= 0.3$ for cooling

For large temperature differences, E. N. Sieder and G. E. Tate (1) modified the Dittus–Boelter equation to account for changing viscosity with temperature.

$$\mathrm{Nu} = 0.023\mathrm{Re}^{0.8}\,\mathrm{Pr}^{1/3}\left(\frac{\mu}{\mu_w}\right)^{0.14} \tag{6.53}$$

Based on the similarity between the rate processes of heat and momentum transfer, the following correlation, called *Reynolds analogy*, has been developed relating the heat transfer coefficient to the friction factor for gases with $\text{Pr} \approx 1.0$.

$$\text{Nu} = \frac{f}{8}\,\text{Re Pr} \tag{6.54}$$

The Darcy-Weisback friction factor f can be read from the Moody diagram (Figure 3.3).

Turbulent Flow in an Annular Passage Concentric pipes are often used to form a heat exchanger, with one fluid flowing through the center pipe and a second fluid flowing through the annular space between the pipes. The Sieder–Tate equation can be used to find h for the outside surface of the inner pipe, replacing the pipe diameter D with an equivalent diameter D_e for the annulus.

$$D_e = \frac{4 \times \text{flow area}}{\text{wetted perimeter}} = \frac{\pi(D_2^{\,2} - D_1^{\,2})}{\pi(D_2 + D_1)} = D_2 - D_1$$

$$\frac{hD_e}{k} = 0.023\left(\frac{\rho V D_e}{\mu}\right)^{0.8}\left(\frac{c_p\mu}{k}\right)^{1/3}\left(\frac{\mu}{\mu_w}\right)^{0.14} \tag{6.55}$$

Gases Flowing across a Cylinder S. T. Hsu (2) describes the work by M. J. M. Douglas and S. W. Churchill, who derived the following equation:

$$\frac{hD}{k_f} = 0.46\left(\frac{DU_\infty}{\nu_f}\right)^{1/2} + 0.00128\left(\frac{DU_\infty}{\nu_f}\right) \tag{6.56}$$

Note that the fluid properties are evaluated at the film temperature. This equation is valid for $DU_\infty/\nu_f > 500$. For $DU_\infty/\nu_f < 500$, Hsu suggests the following equation:

$$\frac{hD}{k_f} = 0.43 + 0.48\left(\frac{DU_\infty}{\nu_f}\right)^{1/2} \tag{6.57}$$

Flow over Spheres An equation that can be used for flow of liquids and gases over spheres is due to S. Whitaker (6):

$$\text{Nu} = 2 + (0.4\text{Re}^{1/2} + 0.06\text{Re}^{2/3})\text{Pr}^{0.4}\left(\frac{\mu_\infty}{\mu_w}\right)^{1/4} \tag{6.58}$$

The fluid properties are evaluated at the temperature of the free stream. For flow velocities over the sphere approaching zero, the Nusselt number approaches a constant 2, which is the value for pure conduction into stationary infinite surroundings.

Flow over Flat Plates For laminar flow over flat plates ($\text{Re} < 5 \times 10^5$), the average Nusselt number is given by the equation

$$\text{Nu} = 0.664\text{Pr}^{1/3}\,\text{Re}^{1/2} \tag{6.59}$$

Re and Nu are based on L, the length from the leading edge of the plate. The physical properties are evaluated at the film temperature.

For turbulent flow over flat plates ($\text{Re} > 5 \times 10^5$), the following equation can be used:

$$\text{Nu} = 0.037\text{Pr}^{1/3}(\text{Re}^{0.8} - 23{,}100) \tag{6.60}$$

6.8.2 Some Correlations for Natural Convection

Horizontal Cylinders McAdams (1) suggests the following correlation equation for natural convection from a horizontal cylinder.

$$\begin{aligned}\text{Nu} &= 0.53(\text{Gr Pr})^{0.25} \\ &= 0.53\left[\frac{D^3\rho^2 g\beta\,\Delta T}{\mu^2}\left(\frac{c_p\mu}{k}\right)\right]^{0.25}\end{aligned} \tag{6.61}$$

The correlation can be used in the range where the product Gr Pr is between 10^3 and 10^9. The fluid properties are evaluated at the film temperature.

Spheres The following correlation can be used for natural convection from spheres:

$$\text{Nu} = 2 + 0.45(\text{Gr Pr})^{1/4} \tag{6.62}$$

Again, the number 2 relates to conduction into a stationary infinite medium.

Effect of Radiation Heat Transfer When evaluating heat transfer at moderate temperatures with forced convection, radiation effects are small by comparison and may be ignored. With natural convection, the convective surface coefficients are relatively small, and the energy transfer by radiation may be significant by comparison. Radiation is treated in some detail in Chapter 8, but for the present we will note that radiant heat transfer from a small object, such as a sphere or pipe, to surrounding walls can be described by the following equation.

$$\frac{q_r}{A} = \varepsilon\sigma(T_1^4 - T_2^4) \tag{6.63}$$

where q_r = net rate of radiant heat transfer from pipe or sphere to surrounding walls

A = surface area of pipe or sphere

ε = emissivity of the small body surface

σ = Stefan–Boltzmann constant = 5.669×10^{-8} W/m^2-K^4

T_1, T_2 = Kelvin

From equation (6.63) we can derive an approximate radiation surface coefficient as follows:

$$q_r = h_r A(T_1 - T_2) \tag{6.64}$$

Thus

$$h_r = \frac{q_r/A}{T_1 - T_2}$$

$$= \frac{\varepsilon\sigma(T_1^4 - T_2^4)}{T_1 - T_2}$$

$$= \varepsilon\sigma(T_1 + T_2)(T_1^2 + T_2^2)$$

or

$$h_r \approx 4\varepsilon\sigma(T_{\text{ave}})^3 \tag{6.65}$$

The overall heat loss from a pipe or sphere in natural convection can be found after evaluating h_c by equation (6.61) or (6.62) and h_r from (6.65).

$$q_{\text{overall}} = (h_c + h_r)A(T_w - T_\infty) \tag{6.66}$$

Natural Convection on Vertical Planes In the laminar range, with the product (Gr Pr) between 10^4 and 10^9, the following equation can be used, with the fluid properties evaluated at the film temperature, and the length is the plate height.

$$\text{Nu} = 0.59(\text{Gr Pr})^{1/3} \tag{6.67}$$

Example 6.8.1

Saturated steam at 200°C flows through a 75-mm OD steel pipe located in a boiler room where the ambient temperature is 30°C. Estimate the rate of heat loss per meter of pipe for the following arrangements:
(a) The pipe is bare.
(b) The pipe is covered with a 25-mm-thick layer of 85% magnesia insulation.
The emissivity of the exposed surface, whether bare steel or insulation, is estimated to be $\varepsilon = 0.6$. The resistance to heat flow of the inside film and metal pipe wall is small and can be ignored. Thus, in either case (a) or (b), the temperature of the outside surface of the steel pipe is approximately 200°C.

Solution

(a) For the bare pipe, the air film properties are evaluated at the average film temperature.

$$T_{\text{film}} = \frac{200 + 30}{2} = 115°\text{C} = 388 \text{ K}$$

$$\nu_{\text{at 388 K}} = 24.6 \times 10^{-6} \text{ m}^2/\text{s}$$

$$\beta = \frac{1}{v}\left(\frac{\partial v}{\partial T}\right)_p = \frac{1}{T} = 2.58 \times 10^{-3} \text{ K}^{-1}$$

$$\Delta T = 200 - 30 = 170 \text{ K}$$

$$\text{Gr} = \frac{D^3 g\beta\, \Delta T}{\nu^2}$$

$$= \frac{(0.075 \text{ m})^3 \times (9.81 \text{ m/s}^2) \times (2.58 \times 10^{-3} \text{ K}^{-1}) \times (170 \text{ K})}{(24.6 \times 10^{-6} \text{ m}^2\text{/s})^2}$$

$$= 3.00 \times 10^6$$

$$\text{Pr} = 0.72$$

$$k_{\text{air at 388 K}} = 0.0313 \text{ W/m-K}$$

Using equation (6.61),

$$\text{Nu} = \frac{h_c D}{k} = 0.53(\text{Gr Pr})^{1/4} = 20.3$$

$$h_c = \frac{20.3 \times 0.0313 \text{ W/m-K}}{0.075 \text{ m}} = 8.47 \text{ W/m}^2\text{-K}$$

Using equation (6.65) for radiation,

$$h_r \approx 4\varepsilon\sigma(T_{\text{ave}})^3$$

$$\approx 4 \times 0.6 \times (5.669 \times 10^{-8} \text{ W/m}^2\text{-K}^4) \times (388 \text{ K})^3$$

$$\approx 7.95 \text{ W/m}^2\text{-K}$$

The overall surface coefficient for the bare pipe is given by

$$h_{cr} = h_c + h_r = 8.47 + 7.95 = 16.4 \text{ W/m}^2\text{-K}$$

Heat loss per meter of bare pipe is

$$q = hA(T_w - T_\infty)$$

$$= (16.4 \text{ W/m}^2\text{-K}) \times (\pi \times 0.075 \times 1 \text{ m}^2) \times (170 \text{ K})$$

$$= 658 \text{ W}$$

(b) Including insulation, there are now two significant resistances to heat transfer, the insulation layer itself and the outside surface air film. Thus heat loss per meter of pipe is given by

$$q = \frac{\Delta T}{\Sigma R} = \frac{\Delta T}{(1/2\pi\bar{k}) \ln (r_0/r_i) + (1/2\pi r_0 h)}$$

To evaluate $\bar{k}$ and h, we must first make a guess at the temperature of the insulation outer surface. After solving, this guess will be checked and improved if necessary. Try $T_{\text{surface}} = 80°\text{C}$.
For 85% magnesia insulation,

$$T_{\text{ave}} = \frac{200 + 80}{2} = 140°\text{C}$$

$$\bar{k} \approx 0.069 \text{ W/m-K}$$

$$R_{\text{insulation}} = \frac{\ln (r_0/r_i)}{2\pi\bar{k}} = \frac{\ln (62.5/37.5)}{2\pi \times 0.069} = 1.18 \text{ K/W}$$

For the outside air film,

$$T_{\text{film}} = \frac{80 + 30}{2} = 55°\text{C} = 328 \text{ K}$$

$$\nu_{\text{air at 328 K}} = 18.4 \times 10^{-6} \text{ m}^2/\text{s}$$

$$\beta = \frac{1}{T} = 3.05 \times 10^{-3} K^{-1}$$

$$\Delta T = 80 - 30 = 50 \text{ K}$$

$$\text{Gr} = \frac{(0.125)^3 \times 9.81 \times (3.05 \times 10^{-3}) \times 50}{(18.4 \times 10^{-6})^2}$$

$$= 8.6 \times 10^6$$

$$\text{Pr} = 0.72$$

$$k_{\text{air at 328 K}} = 0.0277 \text{ W/m-K}$$

$$\text{Nu} = \frac{h_c D}{k} = 0.53(\text{Gr Pr})^{1/4} = 26.4$$

$$h_c = \frac{26.4 \times 0.0277}{0.125} = 5.85 \text{ W/m}^2\text{-K}$$

$$h_r = 4\varepsilon\sigma(T_{\text{ave}})^3$$

$$= 4 \times 0.6 \times (5.669 \times 10^{-8}) \times (328)^3 = 4.80 \text{ W/m}^2\text{-K}$$

$$h_{cr} = h_c + h_r = 10.7 \text{ W/m}^2\text{-K}$$

$$R_{\text{surface}} = \frac{1}{2\pi r_0 h}$$

$$= \frac{1}{2\pi \times (0.125/2) \times 10.7} = 0.238 \text{ K/W}$$

Heat loss per meter of insulated pipe is

$$q = \frac{\Delta T}{\Sigma R} = \frac{200 - 30}{1.18 + 0.238} = 120 \text{ W}$$

The resistances calculated are correct only if the surface temperature is 80°C. We can now make an improved estimate of surface temperature by considering the temperature drop across the air film.

$$q = \frac{T_{\text{surface}} - T_\infty}{R_{\text{surface}}}$$

$$T_{\text{surface}} = T_\infty + q \times R_{\text{surface}}$$

$$= 30 + 120 \times 0.238 = 58.6°\text{C}$$

For our second attempt, try $T_{\text{surface}} = 60°\text{C}$. Now, for an average insulation temperature of 130°C and an air film temperature of 318 K, we correct the property values for the insulation and air film. Repeating all calculations precisely as in our first attempt, we arrive at the following:

$$k_{\text{insulation}} = 0.068$$

$$R_{\text{insulation}} = 1.20$$

$$\nu_{\text{air}} = 17.4 \times 10^{-6}$$

$$\beta = 3.14 \times 10^{-3}$$

$$\Delta T = 60 - 30 = 30$$

$$\text{Gr} = 5.96 \times 10^{6}$$

$$\text{Pr} = 0.72$$

$$k_{\text{air}} = 0.0271$$

$$\text{Nu} = 24.1$$

$$h_c = 5.23$$

$$h_r = 4.38$$

$$h_{cr} = 9.61$$

$$R_{\text{surface}} = 0.265$$

$$q = \frac{200 - 30}{1.20 + 0.265} = 116 \text{ W}$$

$$T_{\text{surface}} = T_\infty + q \times R_{\text{surface}}$$

$$= 30 + 116\,(0.265) = 60.8°\text{C}$$

Since the surface temperature now matches our second guess, we can consider the problem solved. The rate of heat loss per meter of insulated pipe is 116 W.

REFERENCES

1. McAdams, W. H., *Heat Transmission*, 3rd ed. New York: McGraw-Hill Book Company, Inc., 1954.
2. Hsu, S. T., *Engineering Heat Transfer*. New York: Van Nostrand Reinhold Company, Inc., 1963.
3. Kreith, F., *Principles of Heat Transfer*, 3rd ed. New York: Intext Education Publishers, 1973.
4. Eckert, E. R. G., and Drake, R. M., *Heat and Mass Transfer*. New York: McGraw-Hill Book Company, 1959.
5. Gebhart, B., *Heat Transfer*, 2nd ed. New York: McGraw-Hill Book Company, 1971.
6. Holman, J. P., *Heat Transfer*, 5th ed. New York: McGraw-Hill Book Company, 1981.

PROBLEMS

6.1. Consider steady-state heat loss by pure conduction from a sphere of radius R into infinite stationary surroundings with thermal conductivity k. The temperature of the sphere surface is maintained at T_s (by an electrical heater) and the temperature of the surroundings at infinity is T_∞. Now consider an imaginary spherical surface at $r > R$ and write Fourier's heat conduction equation for this surface. Integrate from $r = R$ to $r = \infty$ to solve for the rate of heat loss from the sphere. Compare this expression with Newton's law of cooling to find a surface coefficient for the sphere. Finally, show that the Nusselt number hD/k is equal to 2 for pure conduction into infinite stationary surroundings.

6.2. Six watts of electrical power is supplied to a resistance heater in a 10-cm-diameter sphere that is suspended in air at 25°C. The surface of the sphere remains steady at 45°C.

(a) What is the average surface coefficient of heat transfer for the sphere? *Answer:* $h_{\text{ave}} = 9.55$ W/m²-K

(b) Assuming a sphere surface emissivity $\varepsilon = 0.7$ and that the surrounding walls are at 25°C, estimate the fraction of the surface coefficient that is due to radiation. *Answer:* 49%

6.3. A fan blows air at 10 m/s over the sphere described in Problem 6.2. How much electrical power input is required to maintain the sphere temperature at 45°C? Check the relative magnitude of the radiation effect. *Answer:* 30.9 W; radiation ≈9.4% of total

6.4. Twelve meters of copper pipe, 10-mm ID, 14-mm OD connects a hot-water tank at 60°C to a shower head. Surroundings are at 25°C. When the shower is operating, water flows at a rate of 5×10^{-4} m³/s.

(a) Find the rate of heat loss from the 12-m pipe while the water is flowing. Assume the whole pipe surface is at 60°C and that $\varepsilon \simeq 0.2$. *Answer:* 201 W

(b) Estimate the temperature drop of the water between the hot tank and the shower head. *Answer:* 0.097°C

6.5. After the shower in Problem 6.4 is turned off, how long does it take for the water in the pipe to cool to 30°C? (Assume that the only resistance to heat transfer is at the outside surface of the pipe, and that the surface coefficient h remains constant at the value found for part a of Problem 6.4.) *Answer:* 41 min

6.6. Oil flows in laminar fashion through a tube with inside radius R. The velocity profile is described by the equation $V_r/V_{\text{CL}} = 1 - (r/R)^2$. The temperature profile will be assumed:

$$\frac{\theta_r}{\theta_{\text{CL}}} = \frac{T_r - T_w}{T_{\text{CL}} - T_w} = 1 - \frac{9}{5}\left(\frac{r}{R}\right)^2 + \frac{4}{5}\left(\frac{r}{R}\right)^3$$

Calculate the bulk (cup mixing) temperature of the oil using the specified velocity and temperature profiles. For convenience, use T_w as your zero datum for enthalpy (energy) so that the total rate of flow of enthalpy across a section is $\dot{m}c_p\theta_{\text{bulk}}$. Now consider the rate of enthalpy flow through a ring element of width dr, integrate, and equate to the total. *Answer:* $\theta_{\text{bulk}} = (102/175)\theta_{\text{CL}}$

6.7. Continuing from Problem 6.6, find the local surface coefficient for heat transfer from the oil to the tube wall. Note that the temperature difference used in Newton's law of cooling is $T_{\text{bulk}} - T_w$ or just θ_{bulk}. *Answer:* $h = (70/17)(k/D)$

6.8. Continuing from Problem 6.7, find the local rate of change of fluid bulk temperature with respect to distance in the flow direction. (Energy conservation requires that the increase of fluid enthalpy plus the heat loss to the tube wall over a distance Δx measured in the flow direction should sum up to zero.) *Answer:* $(dT_{\text{bulk}}/dx) = -(280/17)\,(k\,\theta_{\text{bulk}}/\rho D^2\,V_{\text{ave}}Cp)$

6.9. Continuing from Problem 6.8 and assuming a constant tube wall temperature, what tube length is required for the temperature difference θ_{bulk} to be halved? *Answer:* $L = (17/280)\,(\rho D^2 V_{\text{ave}} c_p/k)\ln 2$

6.10. Lubricating oil flows through a 10-mm-ID tube at the rate of 0.10 kg/s. The oil temperature falls from 140° to 120°C while the tube is held at a constant 100°C. Use the result of Problem 6.9 to calculate the required tube length. Properties of the oil at the arithmetic mean bulk temperature are as follows:

$$\rho = 820\ \text{kg/m}^3, \qquad \mu = 0.0082\ \text{kg/m-s}, \qquad k = 1.34 \times 10^{-4}\ \text{kW/m-K}$$

$$c_p = 2.35\ \text{kJ/kg-K}$$

Answer: $L \simeq 94$ m

6.11. For comparison, repeat Problem 6.10 using the empirical correlation equation (6.51) for laminar flow in pipes. For the lubricating oil at 100°C, $\mu_w = 0.017$ kg/m-s. Because the unknown tube length is included in the correlation equation, a trial-and-error method will be required. *Answer:* $L \simeq 70$ m

6.12. Air at 25°C, 1 atm pressure, and with $U_\infty = 20$ m/s blows across a 50-mm-diameter cylinder that is at 90°C. What is the rate of heat loss per meter of cylinder? *Answer:* 1000 W/m

6.13. Air at 25°C, 1 atm pressure, and with $U_\infty = 20$ m/s blows smoothly over both sides of a long thin metal strip at 90°C whose width (parallel to the flow) is chosen so that the exposed surface area per meter of length is the same as for the cylinder in Problem 6.12. What is the rate of heat loss per meter of metal strip? *Answer:* 607 W

6.14. Atmospheric air flowing at the rate of 0.4 kg/s is to be heated from 20° to 60°C by passing it through a smooth 140-mm tube that is kept at 100°C.

(a) Compare the values of h predicted by correlations (6.52), (6.53), and (6.54). *Answers:* 64; 63; 50 W/m²-K

(b) Find the required tube length. To use h, you will need the logarithmic mean temperature difference,

$$\Delta T_m = \frac{\Delta T_1 - \Delta T_2}{\ln(\Delta T_1/\Delta T_2)}$$

where $\Delta T_1 = T_{\text{air in}} - T_w$ and $\Delta T_2 = T_{\text{air out}} - T_w$. *Answer:* 10 m

6.15. A hot fluid flowing in a tube transfers heat to cooler surroundings. Starting with a second-degree polynomial (as follows),

$$\frac{\theta_r}{\theta_{CL}} = \frac{T_r - T_w}{T_{CL} - T_w} = a + b\left(\frac{r}{R}\right) + c\left(\frac{r}{R}\right)^2$$

develop an expression for the temperature profile in the fluid. Three boundary conditions are to be satisfied.

(1) $T_r = T_{CL}$ at $r = 0$
(2) $T_r = T_w$ at $r = R$
(3) $dT_r/dr = 0$ at $r = 0$

Answer: $\theta_r/\theta_{CL} = 1 - (r/R)^2$

6.16. In Problem 6.6, involving fluid flow in a tube with heat transfer, we assumed a temperature profile given by the following expression:

$$\frac{\theta_r}{\theta_{CL}} = \frac{T_r - T_w}{T_{CL} - T_w} = 1 - \frac{9}{5}\left(\frac{r}{R}\right)^2 + \frac{4}{5}\left(\frac{r}{R}\right)^3$$

Boundary conditions satisfied include the three specified in Problem 6.15 and a fourth:

(4) $\dfrac{d^2T_r}{dr^2} = -\dfrac{1}{R}\dfrac{dT_r}{dr}$ at $r = R$

Starting with a third-degree polynomial and using the four boundary conditions specified, derive the expression for the temperature profile. *Note:* Explanation of fourth boundary condition. Consider a ring-shaped element of fluid with inner radius r, thickness dr, and length Δl. Heat transfer by conduction to the inner surface at r is partly convected away by the fluid motion and partly conducted away from the outer surface. Thus, in general, the radial heat conduction changes with radius. However, in the outermost layer of fluid, the velocity approaches zero, so no energy is convected away by fluid motion. Thus at $r = R$,

$$q_{\text{radial (by conduction)}} = \text{constant}$$

Now

$$q_{\text{radial}} = -kA\frac{dT_r}{dr} = -k(2\pi r\,\Delta l)\frac{dT_r}{dr} = \text{constant} \quad \text{at } r = R$$

$$\frac{dq_{\text{radial}}}{dr} = -2\pi k\,\Delta l\,\frac{d}{dr}\left(r\frac{dT_r}{dr}\right) = 0 \quad \text{at } r = R$$

Thus

$$\frac{d^2T_r}{dr^2} = -\frac{1}{R}\frac{dT_r}{dr} \quad \text{at } r = R$$

Heat Exchangers 7

A visit to a petroleum refinery or a chemical plant will soon reveal how widespread is the use of equipment for transferring heat from one process stream to another. This is done for the purpose of controlling conditions in the manufacturing operations, as well as for saving energy, and is an important factor in the economics of running such industrial plants.

Each exchanger must be carefully designed for the conditions under which it is to be used. The designer must pay attention to factors such as stresses due to pressure and temperature gradients, pressure drop across the equipment, which should be kept as small as possible, and conditions that may cause fouling of the surfaces, requiring that the equipment be frequently shut down and cleaned. Consideration of such factors is beyond the scope of this book.

The discussion here is limited to methods used for predicting performance and for estimating the size and exchanger type required for a particular task.

7.1 Double-Pipe Heat Exchanger

One of the simplest heat exchangers consists of two concentric pipes, as illustrated in Figure 7.1. Heat passes from the hot to the colder fluid through the wall of the inner pipe by the modes of conduction and convection.

For an element of the exchanger of area dA, the bulk temperature of the hot and cold fluid is T_h and T_c, respectively. The temperature profile near the wall of the inner pipe is as indicated in Figure 7.2. The rate of heat transfer through dA is expressed as an overall temperature potential, $T_h - T_c$, divided

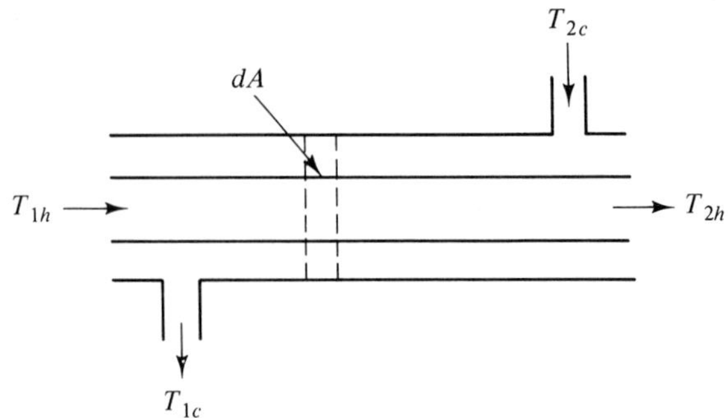

Figure 7.1 Double-pipe heat exchanger

by a total thermal resistance, ΣR, as was done in equation (5.7).

$$Dq = \frac{\Delta T}{\Sigma R} \tag{5.7}$$

where $\Sigma R = 1/U\,dA$

U = overall coefficient of heat transfer

Dq = rate of heat transfer through the area element dA

The total thermal resistance ΣR is the sum of the resistances in the fluids and in the wall.

$$\Sigma R = \frac{1}{U\,dA} = \frac{1}{h_h\,dA_h} + \frac{1}{2\pi k\,dL}\ln\frac{r_c}{r_h} + \frac{1}{h_c\,dA_c} \tag{7.1}$$

where r_c, r_h = outer and inner radii of the inner pipe, respectively

$dA_h = 2\pi r_h\,dL$

$dA_c = 2\pi r_c\,dL$

Figure 7.2 Temperature profile near the exchanger tube wall

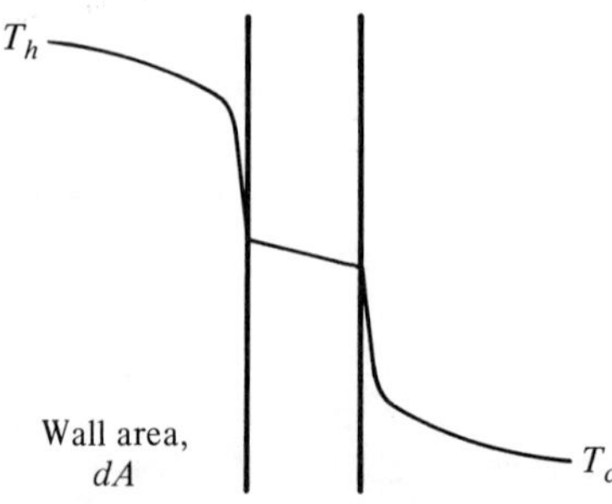

The overall coefficient U can be based on either the inner or the outer surface of the inner pipe, but whichever area is used must be specified. Thus,

$$\Sigma R = \frac{1}{U_h\, dA_h} = \frac{1}{U_c\, dA_c} = \frac{1}{h_h\, dA_h} + \frac{1}{2\pi k\, dL} \ln \frac{r_c}{r_h} + \frac{1}{h_c\, dA_c}$$

and

$$U_h = \frac{1}{(1/h_h) + (r_h/k) \ln (r_c/r_h) + (r_h/h_c r_c)} \tag{7.2}$$

$$U_c = \frac{1}{(r_c/h_h r_h) + (r_c/k) \ln (r_c/r_h) + (1/h_c)} \tag{7.3}$$

Example 7.1.1

A double-pipe heat exchanger is made of 2-in., schedule-40 steel pipe inside 3-in. schedule-40 steel pipe. It is to be used to heat air flowing through the annulus by means of hot water flowing through the inner pipe. The individual coefficients of heat transfer are 85 for the air and 450 for the water (in W/m^2-°C). Calculate the overall coefficient U based on:

(a) Inside area of the inner pipe

(b) Outside area of the inner pipe

The coefficient U is calculated using equation (7.2) or (7.3). The dimensions of the inner pipe are found by referring to a table of standard pipe dimensions. Such tables can be found in handbooks (1, 2). Thus, for the 2-in. pipe,

$$\text{Outside diameter} = 6.033 \text{ cm}$$

$$\text{Inside diameter} = 5.250 \text{ cm}$$

From Table A.3, k for steel ≈45 W/m-°C.

(a) U based on the inner area:

$$\frac{1}{U_i} = \frac{1}{h_i} + \frac{D_i}{2k} \ln \frac{D_o}{D_i} + \frac{D_i}{h_o\, D_o}$$

$$= \frac{1}{450} + \frac{5.250}{2 \times 100 \times 45} \ln \frac{6.033}{5.250} + \frac{5.250}{85 \times 6.033} = 0.0125$$

or $U_i = 80$ W/m^2-°C.

(b) U based on the outer area:

$$\frac{1}{U_o} = \frac{D_o}{h_i D_i} + \frac{D_o}{2k} \ln \frac{D_o}{D_i} + \frac{1}{h_o}$$

$$= \frac{6.033}{450 \times 5.250} + \frac{6.033}{2 \times 100 \times 45} \ln \frac{6.033}{5.250} + \frac{1}{85} = 0.0144$$

or $U_o = 69$ W/m^2-°C.

7.2 Mean Temperature Difference for a Double-Pipe Exchanger

In Figure 7.1, the two streams are shown moving in opposite directions through the heat exchanger. The other arrangement would be for the streams to move in the same direction. Figure 7.3 illustrates the temperature profile for each stream for the two arrangements of the flows. At any cross section of the exchanger, the difference in temperature, ΔT, is the potential for transferring heat from one stream to the other. With the countercurrent arrangement of flows, ΔT is approximately constant, while for the cocurrent arrangement, ΔT varies widely over the length of the exchanger.

For the element of exchanger length indicated in Figure 7.1, the rate of heat transfer is given by equation (5.10).

$$Dq = U\,dA(T_h - T_c) \tag{5.10}$$

If no heat is lost from the exchanger, the heat absorbed by the cold stream must equal that lost by the hot stream over the same area dA. The negative signs in equation (7.4) are consistent with the counterflow arrangement in Figure 7.1.

$$Dq = -\dot{m}_c Cp_c\, dT_c = -\dot{m}_h Cp_h\, dT_h \tag{7.4}$$

where $\dot{m}$ = mass-flow rate

Cp = heat capacity

From equation (7.4),

$$dT_h = -\frac{Dq}{\dot{m}_h Cp_h}, \qquad dT_c = -\frac{Dq}{\dot{m}_c Cp_c}$$

Figure 7.3 Temperature profile in a double-pipe heat exchanger

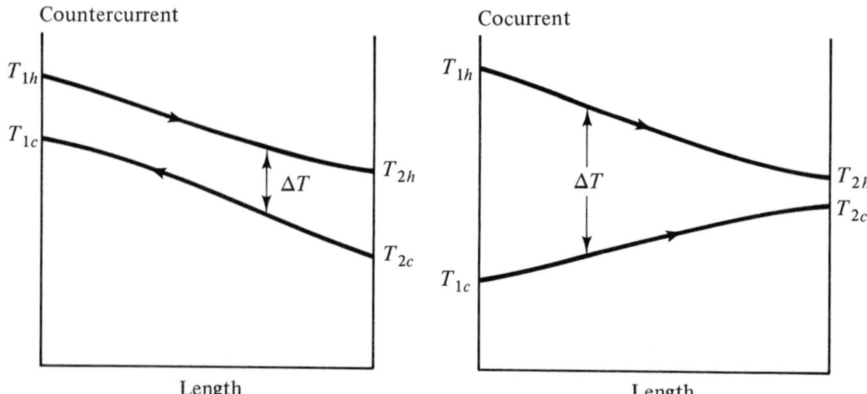

and

$$d(T_h - T_c) = -Dq\left(\frac{1}{\dot{m}_h Cp_h} - \frac{1}{\dot{m}_c Cp_c}\right) \tag{7.5}$$

Substituting for Dq in equation (7.5) from (5.10) gives

$$d(T_h - T_c) = -U(T_h - T_c)\, dA\left(\frac{1}{\dot{m}_h Cp_h} - \frac{1}{\dot{m}_c Cp_c}\right)$$

or

$$\frac{d(T_h - T_c)}{T_h - T_c} = -U\, dA\left(\frac{1}{\dot{m}_h Cp_h} - \frac{1}{\dot{m}_c Cp_c}\right) \tag{7.6}$$

Integrating over the length of the exchanger for steady-state conditions and constant U gives

$$\ln\frac{\Delta T_2}{\Delta T_1} = -UA\left(\frac{1}{\dot{m}_h Cp_h} - \frac{1}{\dot{m}_c Cp_c}\right) \tag{7.7}$$

where $\Delta T_1 = T_{1h} - T_{1c}$

$\Delta T_2 = T_{2h} - T_{2c}$

A heat balance for each stream over the entire exchanger gives

$$q = \dot{m}_h Cp_h(T_{1h} - T_{2h}) = \dot{m}_c Cp_c(T_{1c} - T_{2c}) \tag{7.8}$$

Using equation (7.8) to eliminate the $\dot{m}Cp$ terms from (7.7) results in

$$\ln\frac{\Delta T_2}{\Delta T_1} = -\frac{UA}{q}\left[(T_{1h} - T_{2h}) - (T_{1c} - T_{2c})\right]$$

$$\ln\frac{\Delta T_2}{\Delta T_1} = \frac{UA}{q}(\Delta T_2 - \Delta T_1)$$

or

$$q = UA\,\Delta T_m \tag{7.9}$$

where the mean ΔT is the log-mean temperature difference valid for either parallel or counter flow:

$$\Delta T_m = \frac{\Delta T_2 - \Delta T_1}{\ln(\Delta T_2/\Delta T_1)}$$

The capacity or design equation (7.9) permits the required exchanger area A to be calculated for a specified heat load q and stated values for the terminal temperatures and for a known value of U. It often turns out that the area required cannot reasonably be obtained using two concentric tubes, as indicated in Figure 7.1. In such cases a number of small-diameter tubes are used to give the necessary area A. These are arranged parallel to each other as a bundle and placed inside a large-diameter tube or shell. Such an exchanger is no longer a double-pipe exchanger but is more properly referred to as a multitube exchanger with one tube pass and one shell pass.

7.3 Multipass Exchangers

A multitube exchanger with two tube passes and one shell pass is illustrated in Figure 7.4. Baffles in the shell direct the shell-side fluid across the tubes. The mean ΔT in such an exchanger is not as easily defined as for the double-pipe exchanger. The usual practice is to introduce a correction factor F in the design equation, giving

$$q = UAF\,\Delta T_m \tag{7.10}$$

and to use published charts (3) for evaluating F. In equation (7.10), ΔT_m is the log mean ΔT calculated for the counterflow double-pipe arrangement. Figure 7.5 is a reproduction of one of the correction charts. Others are available for multiple shell passes and for cross-flow heat exchangers.

Example 7.3.1

A shell-and-tube exchanger is to be used to heat benzene by having hot water flow through the single-pass shell. Pressure-drop considerations fix the maximum velocity of the benzene in the tubes at 1.2 m/s. Space limitations require that the exchanger length not exceed 6 m. Determine (a) the number of tube passes, (b) the number of tubes per pass, and (c) the length of the tubes if 13-mm ID tubes are used.

Figure 7.4 Two-tube pass, one-shell pass exchanger

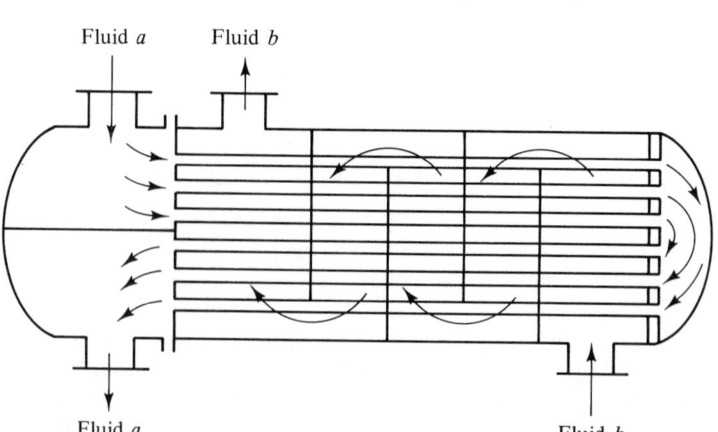

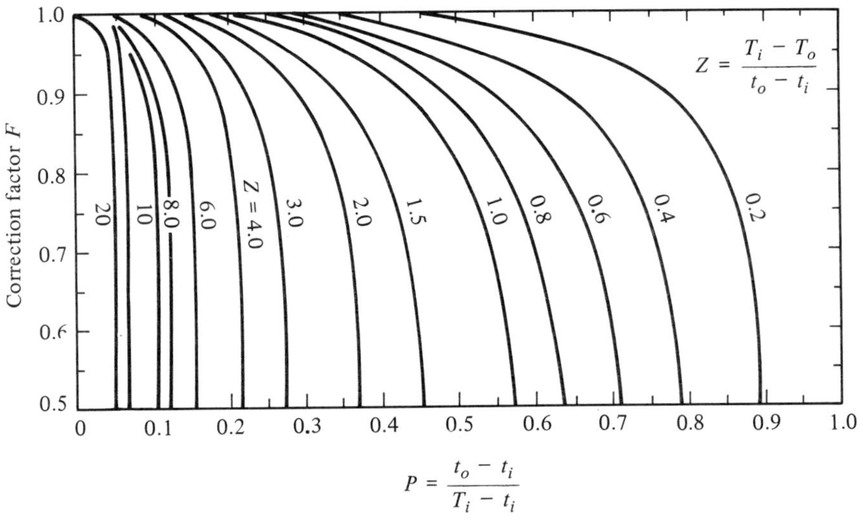

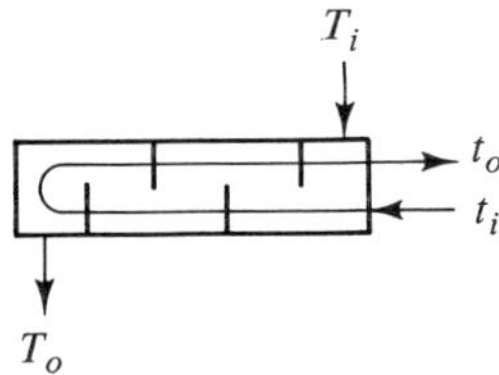

Figure 7.5 Correction factors for heat exchangers with one shell pass and two (or multiples of two) tube passes (3)

Data: Benzene, 6800 kg/hr entering the exchanger at 22°C and leaving at 70°C. Water, 4500 kg/hr entering the exchanger at 100°C. The overall coefficient of heat transfer, U_i, has been estimated to be 284 W/m²-°C. Heat capacity of benzene = 1.88 kJ/kg-°C. Density of benzene = 857 kg/m³.

First Design Calculation

Assume 1 tube pass and determine if this is consistent with the exchanger specifications. The exchanger capacity q is calculated from the heat absorbed by the benzene stream.

$$q = \dot{m}Cp\ \Delta T = 6800\ \frac{\text{kg}}{\text{hr}} \times 1.88\ \text{kJ/kg-°C}\ (70 - 22)\text{°C} \times 1000\ \text{J/kJ} \times \frac{\text{hr}}{3600\ \text{s}}$$

$$= 170{,}453\ \text{W}$$

To calculate the exchanger area A, the exit water temperature T_o and mean temperature difference ΔT_m are required. This is obtained by first writing a heat balance on the exchanger, assuming there are no losses.

$$(\dot{m}Cp\ \Delta T)_{\text{benzene}} = (\dot{m}Cp\ \Delta T)_{\text{water}}$$

Thus

$$(\Delta T)_{\text{water}} = 100 - T_o = \frac{170{,}453}{4500 \times 4.17} \times \frac{3600}{1000} = 32.7°\text{C}$$

$$T_o = 67.3°\text{C}$$

and

$$\Delta T_m = \frac{(67.3 - 22) - (100 - 70)}{\ln (45.3/30)} = 37.1°\text{C}$$

The heat transfer area A_i is now calculated from equation (7.9):

$$A_i = \frac{q}{U_i \, \Delta T_m} = \frac{170{,}453}{284 \times 37.1} = 16.2 \text{ m}^2$$

This area is provided by n tubes in parallel of length L. If the specified maximum velocity for the benzene is used, the value of n can be calculated.

$$n = \frac{\text{total volumetric flow of benzene}}{\text{volumetric flow per tube}}$$

$$= \frac{6800/(3600 \times 857) \text{ m}^3/\text{s}}{1.2(\pi/4)(13 \times 10^{-3})^2 \text{ m}^3/\text{s}} = 13.8 \rightarrow 14 \text{ tubes}$$

Thus

$$A_i = \pi D_i L n$$

and

$$L = \frac{16.2}{\pi \times 13 \times 10^{-3} \times 14} = 28.3 \text{ m}$$

This design is unacceptable since the tube length exceeds the specified length of 6 m.

Second Design Calculation

If a single tube pass exchanger is retained, the tube length may be set at 6 m and the number of tubes increased to obtain the necessary area. Thus

$$n = \frac{16.2}{\pi \times 13 \times 10^{-3} \times 6} = 66.1 \rightarrow 66 \text{ tubes}$$

The velocity in each tube is now reduced to the following value:

$$V = \frac{6800/3600 \times 857}{66(\pi/4)(13 \times 10^{-3})^2} = 0.25 \text{ m/s}$$

This reduced velocity may reduce the value of the coefficient U_i. Otherwise, this design meets the specifications and is therefore acceptable.

Third Design Calculation

A third possible design is to use two tube passes with the maximum benzene velocity so that there are 14 tubes in each pass. The values of q and ΔT_m are unchanged, but the capacity equation must include the correction factor F. Thus

$q = U_i A_i F \, \Delta T_m$. Use Figure 7.5 to find F.

$$Z = \frac{100 - 67.3}{70 - 22} = 0.68, \qquad P = \frac{70 - 22}{100 - 22} = 0.62$$

and

$$F = 0.76$$

The required heat transfer area is now

$$A_i = \frac{170{,}453}{284 \times 0.76 \times 37.1} = 21.3 \text{ m}^2$$

With 14 tubes per pass, the tube length is

$$L = \frac{21.3}{2 \times 14\pi \times 13 \times 10^{-3}} = 18.6 \text{ m}$$

This design is also unacceptable since the tube length exceeds 6 m.

Fourth Design Calculation

A fourth possibility for this exchanger is to retain the two tube passes, set the length of each tube at 6 m and increase the number of tubes so that the required area is obtained. Thus, since $A_i = 21.3$ m^2,

$$n = \frac{21.3}{\pi \times 13 \times 10^{-3} \times 6} = 86.9 \rightarrow 86 \text{ tubes or } 43 \text{ tubes/pass}$$

The benzene velocity through each tube will now be less than the specified maximum.

$$V = \frac{6800/(3600 \times 857)}{(\pi/4)(13 \times 10^{-3})^2 43} = 0.39 \text{ m/s}$$

Fifth Design Calculation

Still another design possibility is to increase the number of tube passes to four. If the benzene velocity in each tube is kept at the maximum of 1.2 m/s, the number of tubes per pass will again be 14. Since q, U_i, F, and ΔT_m are unchanged from design 3, the total required area is $A_i = 21.3$ m^2. The tube length is

$$L = \frac{21.3}{4 \times 14\pi(13 \times 10^{-3})} = 9.31 \text{ m}$$

This is still longer than the specified 6 m and, therefore, this design is also rejected.

Sixth Design Calculation

A final possibility is to retain four tube passes, set the tube length at 6 m, and increase the number of tubes. Thus

$$n = \frac{21.3}{\pi(13 \times 10^{-3})6} = 86.9 \rightarrow 88 \text{ tubes or } 22 \text{ tubes/pass}$$

SUMMARY

In this problem a total of six designs has been considered, beginning with the simplest design of a one-tube pass, one-shell pass exchanger. The results are summarized in Table 7.1.

TABLE 7.1
DESIGN POSSIBILITIES FOR A ONE-SHELL PASS EXCHANGER

Design no.	Tube passes	Tubes per pass	Tube length *m*	Remarks
1	1	14	28.3	Unacceptable
2	1	66	6	Acceptable
3	2	14	18.6	Unacceptable
4	2	43	6	Acceptable
5	4	14	9.31	Unacceptable
6	4	22	6	Acceptable

7.4 Prediction of the Overall Coefficient *U*

In Chapter 6, a few correlations of the many that are available in the literature were presented for predicting the value of the individual heat-transfer coefficient h. In Example 7.1.1, we illustrated how the values of h are combined to give the value of the overall coefficient U. It frequently happens, however, with complex industrial fluids that information required to predict h is not available. In such instances it is useful to have approximate values of U based on someone's experience with similar installations. Table 7.2 lists some values of U for conditions that occur frequently in industry. Other values of U can be found in handbooks such as reference 1.

TABLE 7.2
APPROXIMATE VALUES FOR THE OVERALL COEFFICIENT *U*

Service	U(W/m²-°C)
Steam to:	
Aqueous solutions	550–3300
Heavy fuel oils	50–150
Gases	25–250
Water	1000–3400
Water to:	
Aqueous solutions	550–1100
Organic solvents	280–850
Water	800–1600

7.5 Fouling of Heat-Transfer Surfaces

Heat exchangers in service build up a deposit on the heat-transfer surface that gradually reduces the capacity of the exchanger to transfer heat. The thermal resistance of such deposits is most readily taken into account by adding the appropriate *fouling factors* to equation (7.1) when determining the value of the overall coefficient U. Table 7.3 lists some fouling coefficients. A more extensive list is available in handbooks (1) and reference texts on heat transfer.

TABLE 7.3
HEAT-TRANSFER COEFFICIENT FOR DEPOSITS OR FOULING FACTORS

Service	h (deposit), W/m²-°C
Sea water above 50°C	5600
Brackish water above 50°C	2800
Condensing organic vapors	5600
Process liquids	5600
Boiling liquids	2800
Vegetable oils	1800
Asphalt	560

REFERENCES

1. Perry, R. H., and C. H. Chilton, *Chemical Engineer's Handbook*, 5th ed. New York: McGraw-Hill Book Company, 1973.
2. Parrish, A., *Mechanical Engineer's Reference Book*, 11th ed. London: Butterworth and Co. Ltd., 1973.
3. Bowman, R. A., A. C. Mueller, and W. M. Nagle, "Mean Temperature Difference in Design," *Trans. Am. Soc. Mech. Engrs.*, *62*, 283–294 (1940).

PROBLEMS

7.1. Hot gas at 200°C is fed to a heat exchanger that is so long that it may be considered infinite in length. Cool air at 15°C is passed through the exchanger countercurrent to the hot gas. Calculate the lowest temperature to which the hot gas can be cooled for each of the following cases:
(a) The $\dot{m}C_p$ product for the hot gas is twice that for the air.
(b) The $\dot{m}C_p$ product for the hot gas is equal to that for the air.
(c) The $\dot{m}C_p$ product for the hot gas is one-half that for the air. *Answer:* (a) 107.5°C; (b) 15°C; (c) 15°C

7.2. A double-pipe heat exchanger containing 1.50 m² of surface area is to be used to cool a lubricating oil using water as the coolant. The water is available at 38°C at the rate of 230 kg/hr. The initial temperature of the oil is 120°C. The exit temperature of the two flows must be no greater than 100°C in the case of the water and not less than 60°C for the oil. The overall coefficient of heat transfer has been estimated to be 312 W/m²-°C. Take the C_p value for oil as 2.14 kJ/kg-°C. Determine the maximum flow of oil that may be cooled using this exchanger for (a) countercurrent flow, and (b) cocurrent flow. *Answer:* (a) 1100 kg/hr; (b) 3100 kg/hr

7.3. In a liquid-to-liquid heat exchanger, the hot oil enters the tubes at 205°C and leaves at 94°C, while the cold oil enters the shell at 38°C and leaves at 94°C. Heat loss is negligible. Assuming U is independent of temperature, what will be the true mean overall temperature difference between hot and cold oil in (a) a counterflow exchanger, and (b) the two-tube pass, one-shell pass exchanger shown in Figure 7.4. What advantage does the latter exchanger have over the former? *Answer:* (a) 80.4°C; (b) 66.7°C

7.4. A single-pass shell and tube exchanger is to be used to cool a stream of oil from 125° to 55°C. The coolant is to be water, passing through the shell, which enters at 21°C and leaves the exchanger at 43°C. The overall coefficient is essentially constant and has a value of 170 W/m²-°C.

(a) For an oil flow of 24 kg/min, determine the total surface area required in the exchanger.

(b) If the exchanger is to be 1.8 m long, how many tubes in parallel, each 1.27-cm OD, are required? Take the specific heat of the oil to be 1.97 kJ/kg-°C. *Answer:* (a) 5.95 m²; (b) 83 tubes

7.5. A double-pipe heat exchanger is made using 2-in. schedule-40 steel pipe (5.250-cm ID, 6.033-cm OD) inside 3-in. schedule-40 pipe. The film coefficients of heat transfer have been estimated to be as follows: annulus fluid, 6000; tubeside fluid, 2000, expressed in W/m²-°C. Estimate the overall coefficient U based on (a) the inside and (b) the outside tube area. *Answer:* (a) 1.4×10^3; (b) 1.2×10^3

7.6. A tubular surface condenser is to be designed to handle 136,000 kg/hr of dry saturated steam condensing at an absolute pressure of 5.08 kPa (saturation temperature = 33.2°C, latent heat of condensation = 2424 kJ/kg). The tubes are to be Admiralty metal (k = 109 W/m-°C) with outside and inside diameters of 2.540 and 2.291 cm, respectively. Clean cooling water is available at 21°C. It is planned to use a water velocity of 2 m/s and to permit the water temperature to rise 8.5°C. Determine the following:

(a) m³/min of cooling water required

(b) Number of tubes in parallel

(c) Length of each tube

Values for the individual coefficients of heat transfer, in W/m²-°C, are as follows: water film, 7660; steam film, 17,000; scale deposit, 17,000 on both the water and steam sides. *Answer:* (a) 155 m³/min.; (b) 3133; (c) 17.3 m

7.7. If 19,000 kg/hr of pressurized water is to be preheated from 70° to 180°C in an economizer, and 45,000 kg/hr of flue gas at 370°C is available as the source of heat, determine the area required in the exchanger and the exit temperature of the flue gas for a single-pass, counterflow design. The overall coefficient has been estimated to be 85 W/m²-°C. The heat capacity of the flue gas is 1.05 kJ/kg-°C. *Answer:* 192 m²; 185°C

7.8. A tank with a capacity of 230 kg of liquid plasticizer is to be used to heat the plasticizer from 15° to 85°C. The tank is equipped with a 15-m coil of 2.50-cm OD copper tubing, which is fed with steam having a saturation temperature of 104°C. The tank is also equipped with an electrically driven mixer to provide vigorous agitation. Two methods of operation have been proposed.
Method A: The tank is filled with 230 kg of plasticizer at 15°C and the mixer and steam are turned on. When the temperature of the liquid is 85°C, the mixer and steam are turned off and the tank is drained. The cycle is then repeated. The time required to fill the tank is 5 min and to drain the tank also requires 5 min.
Method B: Plasticizer at 15°C is fed continuously to the tank and withdrawn continuously at 85°C. The mixer and steam are on continuously.
Data: With the mixer on, the overall coefficient of heat transfer U_o, based on the outside area of the coil, is 560 W/m²-°C. Heat capacity of the plasticizer is 2.9 kJ/kg-°C.
Determine the number of kilograms of plasticizer that can be heated per min by each method. *Answer:* 6.4 kg/min by method A; 3.7 kg/min by method B

Radiant Heat Transfer 8

8.1 Introduction

In addition to conduction and convection, a third mode of heat transfer, radiation, becomes increasingly important at high temperatures. Hot surfaces emit energy in the form of electromagnetic waves that travel through space at the speed of light.

$$C = \lambda\nu = 3 \times 10^{10} \text{ cm/s} \tag{8.1}$$

where C = speed of light

λ = radiation wavelength

ν = radiation frequency

Electromagnetic radiation includes a broad spectrum of wavelengths and frequencies, as illustrated in Figure 8.1.

Planck's quantum theory postulates that radiant energy is transmitted in the form of discrete quanta, each of which has an energy

$$E = h\nu \tag{8.2}$$

where h is a constant called *Planck's constant*. This implies that a quantum of radiation at high frequency (short wavelength) is very energetic. However, the energy quanta emitted from hot surfaces are most numerous in the range of the spectrum labeled *thermal*, and it is within this range that most of the thermal radiation energy is found. The actual radiant energy emitted by a hot surface depends on the temperature and emitting properties of the surface.

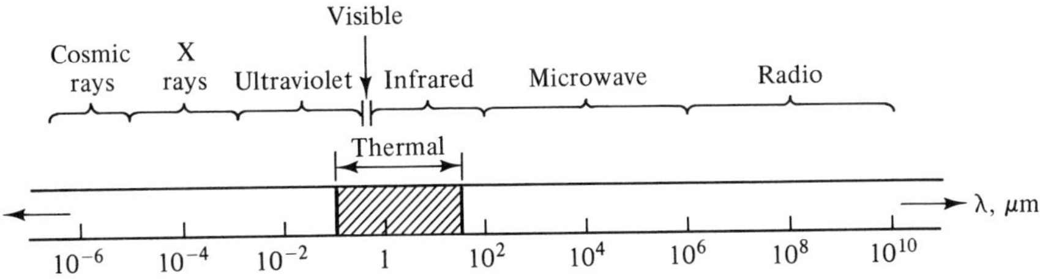

Figure 8.1 Radiation spectrum

8.2 Some Definitions and Properties

Radiation incident on a body may be reflected, absorbed, or transmitted.

$G \equiv$ radiant energy falling on body = irradiation

$\alpha \equiv$ fraction of irradiation absorbed = absorptivity

$\rho \equiv$ fraction of irradiation reflected = reflectivity

$\tau \equiv$ fraction of irradiation transmitted = transmissivity

With these definitions, an energy balance on the body shown in Figure 8.2 gives $G = \alpha G + \rho G + \tau G$. Thus

$$\alpha + \rho + \tau = 1 \tag{8.3}$$

Most solids used in engineering applications are opaque. Thus the transmissivity τ is generally zero, and we have

$$\alpha + \rho = 1 \tag{8.3a}$$

Reflection from a polished surface may be *specular*, as shown in Figure 8.3(a), and from a rough or granular surface is generally considered *diffuse*, as in Figure 8.3(b). For diffuse reflection, the amount of energy reflected in direction ϕ is proportional to $\cos \phi$.

Figure 8.2 Radiation beam incident on a solid surface

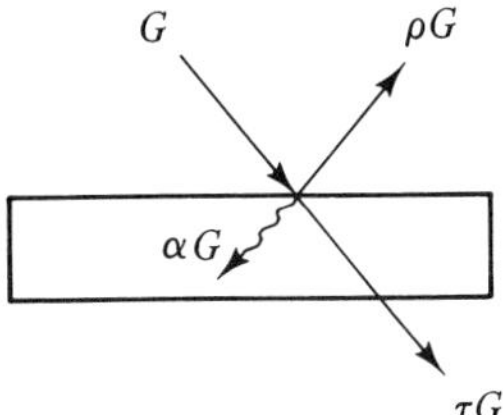

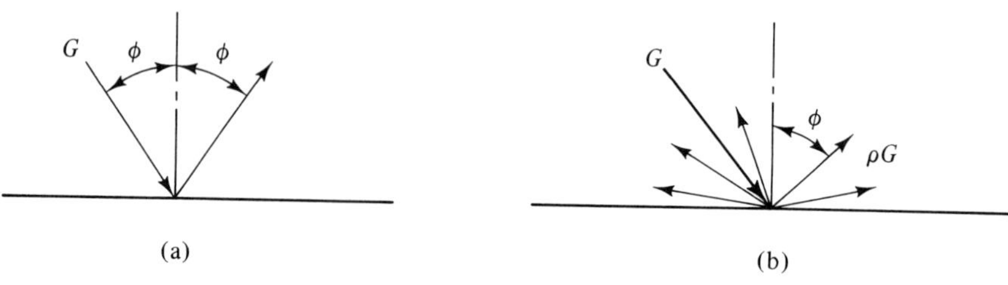

Figure 8.3 (a) Specular reflection from a polished surface; (b) diffuse reflection from a rough, granular surface

8.3 Black-Body Radiation

A real body at some temperature radiates energy. If its surface were modified to make it a better radiator, it would radiate more energy. Is there an upper limit to the amount of energy that may be radiated for a given temperature?

Consider another approach. Radiation impinges on a surface and part is reflected and part absorbed. By altering the surface characteristics, we could increase the absorption and reduce the reflection. A *black body* is defined as one that would absorb all incident radiation, reflecting none.

No real black bodies exist, but a black body can be approximated by a cavity with a small entrance (Figure 8.4). Consider a ray of radiation G_0 entering the cavity. At the first reflection, the amount of energy reflected will be G_1, where

$$G_1 = \rho G_0 = (1 - \alpha)\, G_0$$

When G_1 is again reflected, we have

$$G_2 = \rho G_1 = \rho^2 G_0 = (1 - \alpha)^2\, G_0$$

and after many reflections

$$G_n = (1 - \alpha)^n G_0$$

If n is large before the reflected ray G_n finds the opening to leave, $(1 - \alpha)^n \approx 0$, and thus the amount of G_0 that is reflected from the opening approaches zero.

Figure 8.4 Approximation to a black body

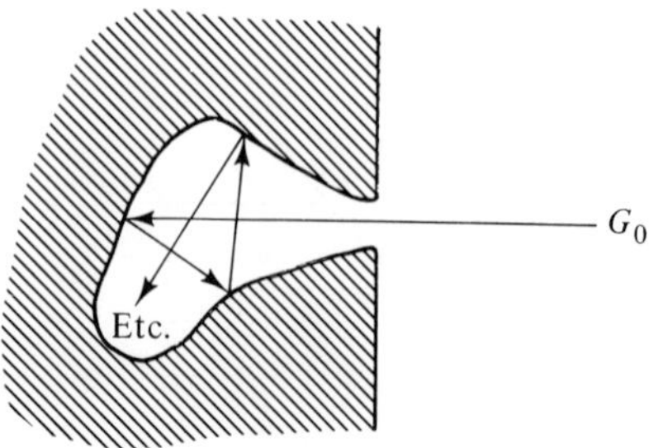

Besides being able to absorb all incident radiation, a black body is also the best possible radiator. If this cavity (wall) was maintained at uniform temperature, the radiation emitted from the hole would be called *black-body radiation* and would be a maximum for this temperature. The total energy emitted per unit area per unit time is called the emissive power, E W/m^2. For a black body, Stefan found experimentally and Boltzmann proved analytically that

$$E_b = \sigma T^4 \tag{8.4}$$

where E_b = black-body emissive power, W/m^2

σ = Stefan–Boltzmann constant = 5.669×10^{-8} W/m^2-K^4

T = K

The emissive power E for a nonblack surface is less than E_b, and the ratio is the total *emissivity*, ε.

$$\varepsilon = \frac{E}{E_b} \tag{8.5}$$

Numerical values of total emissivity are listed in handbooks (1, 2).

8.4 Kirchhoff's Law

A body that is at the same temperature as its surroundings has a total emissivity ε equal to its absorptivity α. To prove this statement, consider a black-body enclosure surrounding a small body, as in Figure 8.5. First assume that the small body is black. The energy radiated from the small black body per unit time to the surrounding enclosure is AE_b, where A is the surface area of the small body. Assuming that the whole system is in thermal equilibrium, the radiation from the surroundings falling on the small body must also equal

Figure 8.5 Enclosed small body

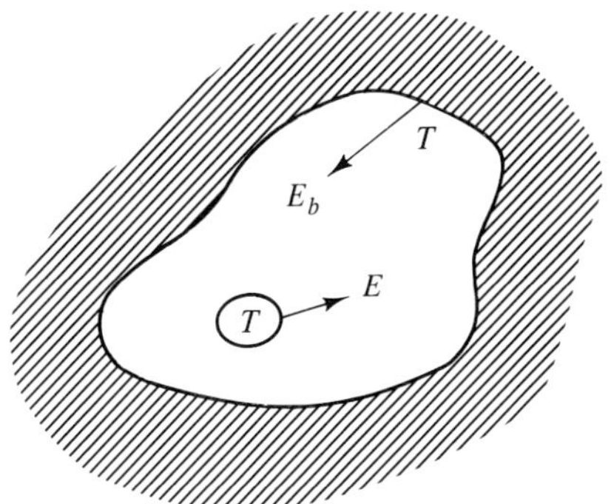

AE_b; otherwise, there would be a net gain or loss of energy by the small body and a corresponding temperature change. This would constitute a violation of the second law of thermodynamics.

Now, replace the small black body with a nonblack body identical in size and shape. The radiation falling on this small nonblack body must be the same as fell on the small black body (AE_b), because the geometry is the same. The energy absorbed by the nonblack body is αAE_b. Again, for thermal equilibrium the energy emitted by the nonblack body must equal the energy absorbed.

$$AE = \alpha AE_b$$

$$\alpha = \frac{E}{E_b} = \varepsilon \tag{8.6}$$

8.5 Monochromatic Emissive Power and Spectral Energy Distribution

The total emissive power E of a hot surface is distributed over the full spectrum of wavelengths. The monochromatic emissive power E_λ is defined by the following expression:

$$E = \int_0^\infty E_\lambda \, d\lambda \tag{8.7}$$

For a radiating surface at a particular temperature T, a plot of E_λ against λ is called the *spectral energy distribution*, and the area under the curve is the total emissive power E at that temperature.

Planck's equation for the monochromatic emissive power of a black body is

$$E_{b\lambda} = \frac{C_1 \lambda^{-5}}{e^{C_2/\lambda T} - 1} \tag{8.8}$$

where λ = wavelength in micrometers (μm)

$C_1 = 3.743 \times 10^8$ W-μm^4/m^2 (1.187×10^8 Btu-μm^4/hr-ft^2)

$C_2 = 1.4387 \times 10^4$ μm-K (2.5896×10^4 μm-°R)

From this equation, curves of monochromatic emissive power can be drawn against λ for different temperatures. We note from Figure 8.6 that the peak in monochromatic emissive power shifts to lower wavelengths with increasing temperature. This shift is described mathematically by Wien's displacement law.

$$\lambda_m T = 2987.6 \text{ } \mu\text{m-K } (5215.6 \text{ } \mu\text{m-°R}) \tag{8.9}$$

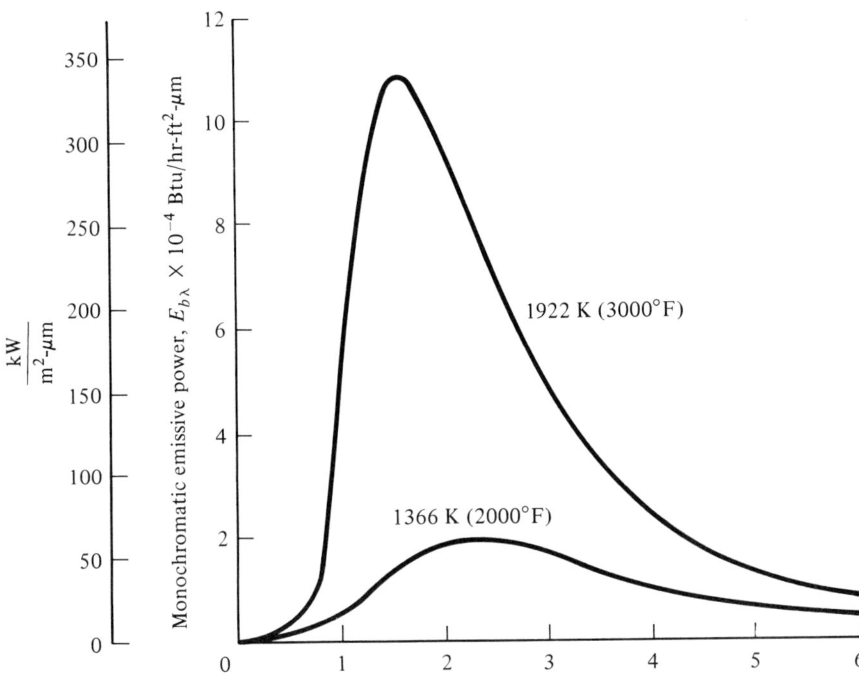

Figure 8.6 Black-body spectral distribution curves

where λ_m = wavelength at which maximum monochromatic emission occurs (μm)

T = absolute temperature of the radiating surface (K or °R)

Some bodies absorb radiation selectively (over specific ranges of wavelength). At a given black-body temperature T, the energy radiated in the wavelength range from 0 to λ is

$$E_{b(0-\lambda)} = \int_0^{\lambda} E_{b\lambda} d\lambda = \int_0^{\lambda T} \frac{1}{T} E_{b\lambda}\, d(\lambda T) \tag{8.10}$$

The fraction of the total black-body radiation E_b in the range 0 to λ is

$$\begin{aligned} \frac{E_{b(0-\lambda)}}{E_b} &= \int_0^{\lambda T} \frac{(1/T)\, E_{b\lambda}\, d(\lambda T)}{\sigma T^4} \\ &= \int_0^{\lambda T} \frac{C_1 \lambda^{-5}}{(e^{C_2/\lambda T} - 1)\sigma T^5}\, d(\lambda T) \qquad (8.11) \\ &= \int_0^{\lambda T} \frac{(C_1/\sigma) d(\lambda T)}{(\lambda T)^5 (e^{C_2/\lambda T} - 1)} \end{aligned}$$

This integral has been solved and tabulated over a wide range of values for λT (see Table A.4). For a black body at temperature T, we can calculate the emissive power in the wavelength range between λ_1 and λ_2 as follows:

$$E_b(\lambda_1 - \lambda_2) = \left[\frac{E_b(0 - \lambda_2 T)}{\sigma T^4} - \frac{E_b(0 - \lambda_1 T)}{\sigma T^4}\right] \sigma T^4 \tag{8.12}$$

The quantities within the brackets [ ] are tabulated in Table A.4 opposite the particular values of $\lambda_2 T$ and $\lambda_1 T$.

8.6 Radiant Heat Transfer between Black Bodies

8.6.1 Two Parallel Black Surfaces. The simplest case to consider is two large black planes separated by a nonabsorbing medium, the first plane at T_1 and the second at T_2 (Figure 8.7). For planes whose dimensions are large relative to the separation distance, the radiation losses at the edges can be ignored. The energy radiated from 1 and absorbed by 2 is equal to $A_1 E_{b_1} = A_1 \sigma T_1^4$. The energy radiated from 2 and absorbed by 1 is equal to $A_2 E_{b_2} = A_2 \sigma T_2^4$. Thus, for planes of equal size ($A_1 = A_2 = A$), the net rate of energy transfer from 1 to 2 is given by

$$q_{\text{net}}(1 \rightarrow 2) = A\sigma(T_1^4 - T_2^4) \tag{8.13}$$

8.6.2 Radiation Power and Net Heat Transfer. As a matter of convenience, we will introduce the symbol R to designate the rate of flow of radiation energy emitted from a surface. Referring to Section 8.6.1, we could replace the cumbersome expression "energy radiated from 1 and absorbed by 2 is equal to $A_1 E_{b1}$" with the algebraic statement $R_{1-2} = A_1 E_{b1}$. Some such symbol is necessary to distinguish between the one-directional flow of radiant energy R_{1-2} and the net energy exchange between two radiating surfaces, q_{net} $(1 \rightarrow 2)$.

$$q_{\text{net}}(1 \rightarrow 2) = R_{1-2} - R_{2-1} \tag{8.14}$$

With this clarification, we will now drop the subscript net from the heat-transfer symbol, so equation (8.13) can be rewritten

$$q_{1-2} = R_{1-2} - R_{2-1} \tag{8.14}$$

Figure 8.7 Radiation between two parallel black surfaces

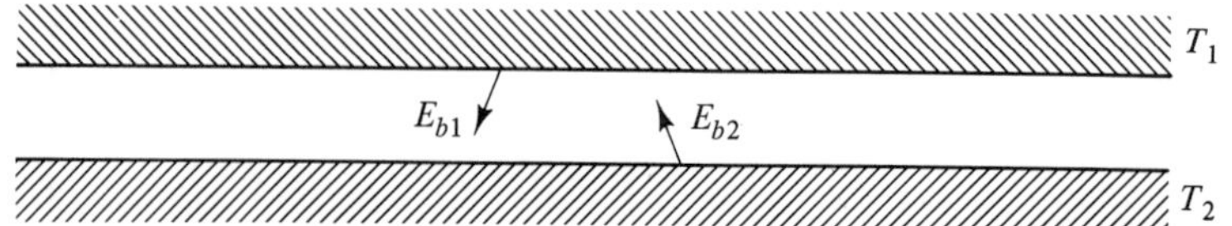

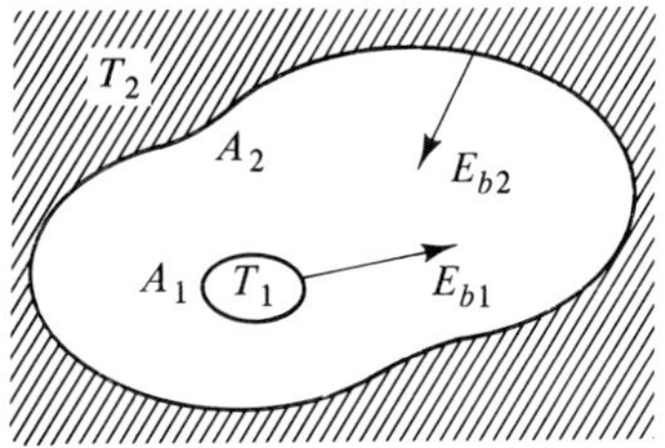

Figure 8.8 Enclosed small body

8.6.3 Small Black Body in Black Enclosure. The energy radiated from 1 to 2 in Figure 8.8 is

$$R_{1-2} = A_1 \sigma T_1^4$$

The energy radiated from 2 to 1 is the same as the energy that would be radiated by 1 at temperature T_2:

$$R_{2-1} = A_1 \sigma T_2^4$$

The net heat transfer from 1 to 2 is

$$q_{1-2} = R_{1-2} - R_{2-1} = A_1 \sigma (T_1^4 - T_2^4) \tag{8.15}$$

8.6.4 Open Systems. If A_2 in Figure 8.8 does not completely surround A_1, then some of the energy emitted by A_1 and A_2 will escape to other surfaces. The net heat transfer from 1 to 2 will then depend on the geometric configuration.

$$q_{1-2} = A_1 F_{1-2} \sigma (T_1^4 - T_2^4) \tag{8.16}$$

F_{1-2} is called the *shape factor* or *view factor* or *configuration factor*. The first step toward the calculation of shape factors is to define radiation intensity.

8.7 Radiation Intensity

Radiation intensity is defined as the rate of flow of radiant energy per unit solid angle per unit of emitter area. The emitter area referred to is the projected area normal to the line of sight of the receiver (i.e., $dA_1 \cos \phi$ in Figure 8.9). The solid angle is defined as $d\omega = dA_2/r^2$. Let R_{1-2} be the rate of flow of radiant energy from dA_1 passing out through the hemisphere. Let dR_{1-2} be the rate of flow of radiant energy from dA_1 to dA_2. If dA_1 in Figure 8.9 is a black body, we have from the definition of radiation intensity

$$I_b = \frac{dR_{1-2}}{(dA_1 \cos \phi)\, d\omega} \tag{8.17}$$

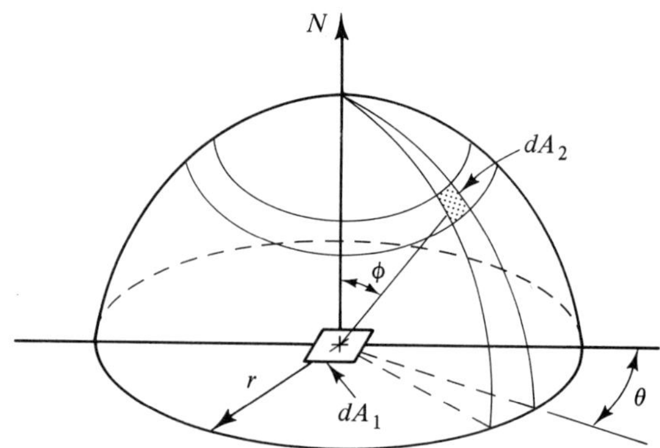

Figure 8.9 Radiation from a point source

Thus the radiant energy flowing from dA_1 to dA_2 is

$$dR_{1-2} = I_b\, dA_1 \cos\phi \frac{dA_2}{r^2} \tag{8.17a}$$

Lambert's law for diffuse radiation states that the energy radiated at angle ϕ is proportional to cosine ϕ [refer to Section 8.2 and Figure 8.3(b)].

$$dR_{1-2} = \text{constant} \times \cos\phi \tag{8.18}$$

Comparing equation (8.17) derived from the definition of intensity and (8.18) expressing Lambert's law, it is apparent that for diffuse radiation (as from a black body) the radiation intensity I_b is constant, independent of the angle ϕ.

We can relate the black-body radiation intensity I_b to the black-body emissive power E_b as follows:

$$dR_{1-2} = I_b\, dA_1 \cos\phi \frac{dA_2}{r^2} \tag{8.17a}$$

Note from Figure 8.9 that $dA_2 = (r\, d\phi)r \sin\phi\, d\theta$. Thus $dR_{1-2} = I_b\, dA_1 \sin\phi \cos\phi\, d\phi\, d\theta$. The emissive power of the black surface dA_1 is defined as

$$\begin{aligned} E_b &= \frac{R_{1-2}}{dA_1} \\ &= \frac{I_b\, dA_1}{dA_1} \int_{\theta=0}^{2\pi} \int_{\phi=0}^{\pi/2} \sin\phi \cos\phi\, d\phi\, d\theta \\ &= I_b \int_0^{2\pi} \left(\frac{\sin^2\phi}{2} \bigg|_0^{\pi/2} \right) d\theta \\ &= \pi I_b \end{aligned} \tag{8.19}$$

Equation (8.19) expresses an important relation. It states that the blackbody emissive power is π times the radiation intensity. It may be confusing to see that the dimensions of I are the same as for E. This is because the unit solid angle (steradian) is dimensionless.

We can now use the concept of intensity to specify the amount of radiation energy emitted from body 1 that impinges on body 2. The radiation striking surface dA_2 is equal to the intensity of radiation emitted from dA_1 times the apparent emitting area (as seen from dA_2) times the solid angle (cone) through which the radiation flows (Figure 8.9).

$$dR_{1-2} = I_{b_1} dA_1 \cos \phi_1 \frac{dA_2}{r^2} \tag{8.20}$$

If dA_2 is not normal to the line of sight (Figures 8.10 and 8.11), then its apparent area (as seen from dA_1) is $dA_2 \cos \phi_2$.

$$dR_{1-2} = I_{b_1}(dA_1 \cos \phi_1) \frac{dA_2 \cos \phi_2}{r^2} \tag{8.21}$$

Figure 8.10

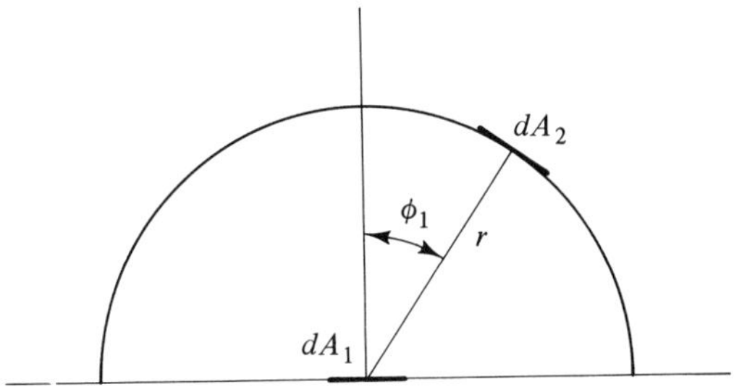

Figure 8.11

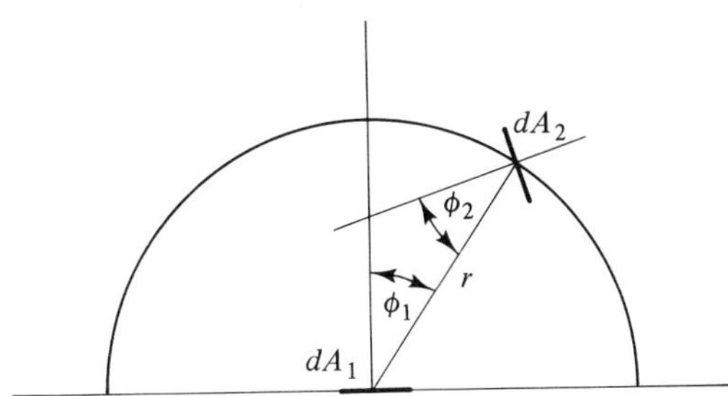

8.8 Geometric Shape Factor, F_{1-2}

The shape factor F_{1-2} is defined as the fraction of the total energy radiated from surface 1 that is intercepted by surface 2 (Figure 8.12). The total energy radiated from dA_1 is

$$R_{dA_1-\text{hemisphere}} = E_{b_1}\, dA_1 = \pi I_{b_1}\, dA_1 \tag{8.22}$$

The amount of this energy intercepted by dA_2 is

$$dR_{dA_1-dA_2} = I_{b_1} \frac{dA_1 \cos \phi_1\, dA_2 \cos \phi_2}{r^2} \tag{8.23}$$

The fraction of the total energy radiated from dA_1 that is intercepted by dA_2 is

$$\begin{aligned} E_{dA_1-dA_2} &= \frac{(I_{b_1}\, dA_1 \cos \phi_1\, dA_2 \cos \phi_2)/r^2}{\pi I_{b_1}\, dA_1} \\ &= \frac{\cos \phi_1 \cos \phi_2\, dA_2}{\pi r^2} \end{aligned} \tag{8.24}$$

If dA_1 emits to a finite surface A_2, equation (8.24) must be integrated over A_2.

$$F_{dA_1-A_2} = \int_{A_2} \frac{\cos \phi_1 \cos \phi_2\, dA_2}{\pi r^2} \tag{8.25}$$

Figure 8.12 Radiation between two surfaces not forming an enclosure

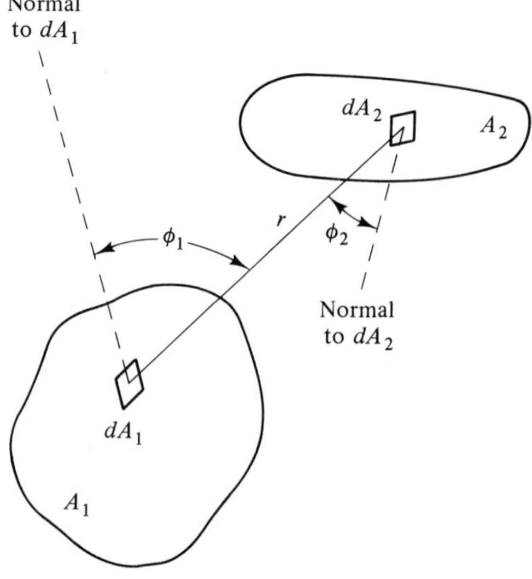

Finally, for radiation from finite A_1 to finite A_2, equation (8.24) must be integrated over both areas A_1 and A_2.

$$F_{A_1-A_2} = \frac{\int_{A_2}\int_{A_1} (I_{b_1}\, dA_1 \cos\phi_1\, dA_2 \cos\phi_2)/r^2}{\int_{A_1} \pi I_{b_1}\, dA_1} \tag{8.26}$$

Now I_{b_1} is constant, so the denominator in equation (8.26) is

$$\int_{A_1} \pi I_{b_1}\, dA_1 = \pi I_{b_1} \int_{A_1} dA_1 = \pi I_{b_1} A_1$$

Also I_{b_1} in the numerator can be removed from the integral signs, since I_{b_1} is constant, and canceled with I_{b_1} in the denominator.
Thus

$$F_{A_1-A_2} = \frac{1}{\pi A_1} \int_{A_2}\int_{A_1} \frac{\cos\phi_1 \cos\phi_2\, dA_1\, dA_2}{r^2} \tag{8.27}$$

Equation (8.27) is seen to be a purely geometric relationship, independent of the radiation properties.

For most practical shapes, this integral (8.27) is cumbersome and difficult. However, it has been solved for various configurations, and the results are available in the form of graphs. Figures A.1 through A.3 in the Appendix present the results for

1. Parallel directly opposed plane circular discs
2. A plane surface radiating to an identical parallel plane
3. Two perpendicular planes with a common edge

Even with these readily available solutions for the shape factor, $F_{A_1-A_2}$, we would be severely restricted in our ability to handle many configurations. Fortunately, a number of relationships between shape factors allow extension of the information given for simple configurations so that it is applicable to more complicated arrangements.

8.9 Shape Factor Relationships

8.9.1 Reciprocity Relation. Rearranging equation (8.27) slightly, we have

$$A_1F_{1-2} = \int_{A_2}\int_{A_1} \frac{\cos\phi_1 \cos\phi_2\, dA_1\, dA_2}{\pi r^2} \tag{8.28}$$

As the subscripts are arbitrary, they can be interchanged to produce the following expression:

$$A_2F_{2-1} = \int_{A_1}\int_{A_2} \frac{\cos \phi_2 \cos \phi_1 \, dA_2 \, dA_1}{\pi r^2} \tag{8.29}$$

Since the double integrals of equations (8.28) and (8.29) differ only in the order of integration, they produce the same results. Therefore,

$$A_1F_{1-2} = A_2F_{2-1} \tag{8.30}$$

8.9.2 Emitter Completely Enclosed by *n* Surfaces

$$\sum_{j=1}^{j=n} F_{i-j} = 1 \tag{8.31}$$

F_{i-j} is the fraction of radiation emitted from surface i that strikes surface j. For a complete enclosure, the sum of the fractions F_{i-j} for the n receiving surfaces must add up to 1.

8.9.3 Subdivision of Receiving Surface

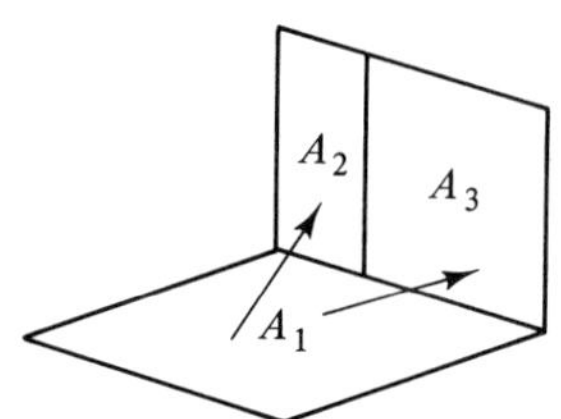

Figure 8.13 Receiver subdivided

$$F_{1-2,3} = F_{1-2} + F_{1-3} \tag{8.32}$$

The fraction of energy emitted by surface A_1 (Figure 8.13) that strikes surface $A_2 + A_3$ is equal to the fraction that strikes A_2 plus the fraction that strikes A_3.

8.9.4 Subdivision of Emitting Surface.

Consider Figure 8.14.

$$(A_1 + A_2)F_{1,2-3} = A_1F_{1-3} + A_2F_{2-3} \tag{8.33}$$

Figure 8.14 Emitter subdivided

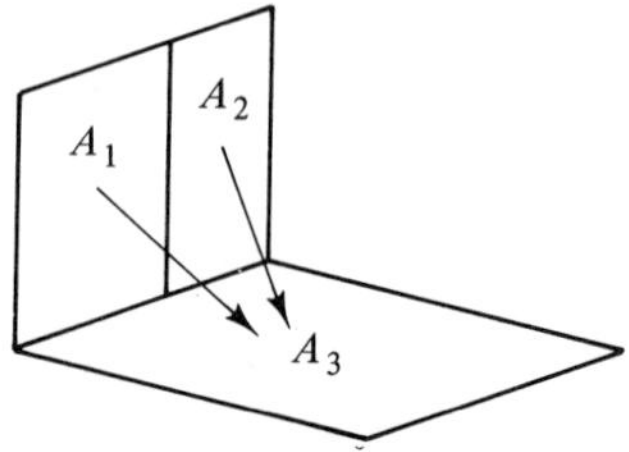

While equation (8.32) should be obvious, (8.33) is not. However, we can use equations (8.30) and (8.32) to derive (8.33). First, use the reciprocity relation (8.30).

$$(A_1 + A_2)F_{1,2-3} = A_3F_{3-1,2}$$

Now use (8.32) to expand $F_{3-1,2}$:

$$(A_1 + A_2)F_{1,2-3} = A_3F_{3-1} + A_3F_{3-2}$$

Finally, use the reciprocity relationship again to give (8.33)

$$(A_1 + A_2)F_{1,2-3} = A_1F_{1-3} + A_2F_{2-3}$$

8.10 Sample Shape Factor Calculations

Example 8.10.1

Find the shape factor F_{1-2} for radiation from area A_1 to area A_2 shown in Figure 8.15. In the Appendix we can find the shape factor for areas such as A_1 and A_2 in Figure 8.16 that have a common edge. A_1 and A_2 in Figure 8.15 do not have a common edge, but by using the shape factor relations we can express F_{1-2} in terms of areas that do have a common edge. Using equation (8.32) for subdivision of the receiving surface,

$$F_{1-23} = F_{1-2} + F_{1-3}$$

Hence

$$F_{1-2} = F_{1-23} - F_{1-3} \tag{8.34}$$

Figure 8.15

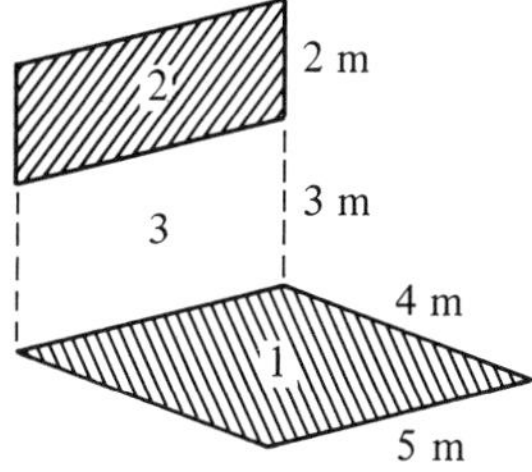

Figure 8.16

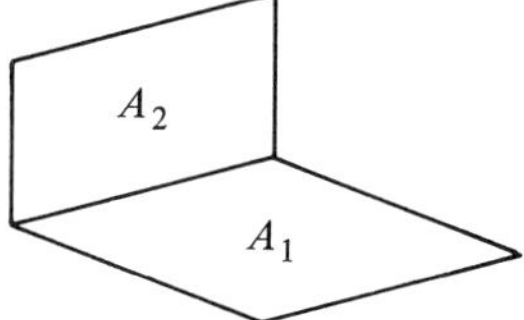

Using Figure A.3, we obtain the following:

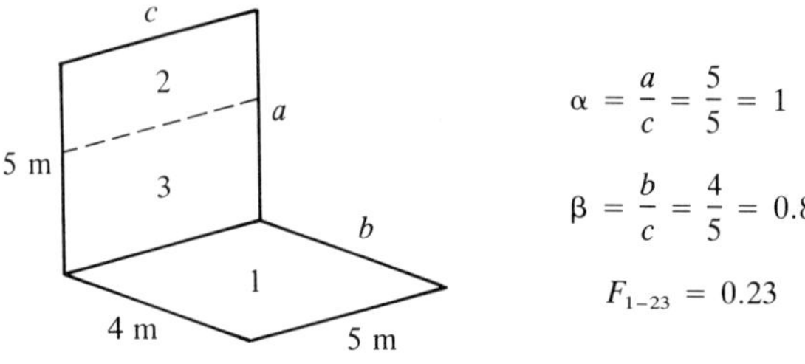

$$\alpha = \frac{a}{c} = \frac{5}{5} = 1$$

$$\beta = \frac{b}{c} = \frac{4}{5} = 0.8$$

$$F_{1-23} = 0.23$$

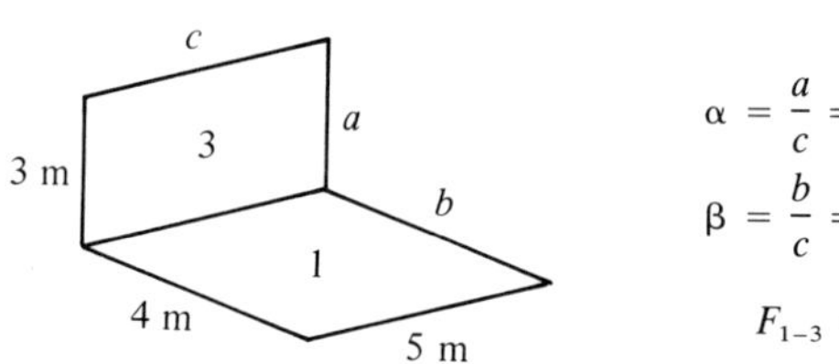

$$\alpha = \frac{a}{c} = \frac{3}{5} = 0.6$$

$$\beta = \frac{b}{c} = \frac{4}{5} = 0.8$$

$$F_{1-3} = 0.20$$

From equation (8.34),

$$F_{1-2} = F_{1-23} - F_{1-3} = 0.23 - 0.20 = 0.03^*$$

Example 8.10.2

Find the shape factor F_{1-2} for radiation from area A_1 to area A_2 shown in Figure 8.17. We transform Figure 8.17 to Figure 8.18 in order to use the data available in Figure A.2 for shape factors for parallel plates. The required shape factor is F_{A1-A2} (Figure 8.17), which is the same as F_{3-56} in Figure 8.18.

$$A_3F_{3-56} = A_{34}F_{34-56} - A_4F_{4-56} \qquad \text{from (8.33)}$$

$$= A_{34}F_{34-56} - A_4F_{4-6} - A_4F_{4-5} \qquad \text{from (8.32)}$$

$$= A_{34}F_{34-56} - A_4F_{4-6} - A_5F_{5-4} \qquad \text{from (8.30)}$$

$$= A_{34}F_{34-56} - A_4F_{4-6} - A_5(F_{5-34} - F_{5-3}) \qquad \text{from (8.32)}$$

By symmetry in Figure (8.18), we have $A_3F_{3-56} = A_5F_{5-34}$. Thus

$$2A_3F_{3-56} = A_{34}F_{34-56} - A_4F_{4-6} + A_5F_{5-3}$$

$$F_{3-56} = \frac{1}{2A_3}(A_{34}F_{34-56} - A_4F_{4-6} + A_5F_{5-3})$$

Using the dimensions from Figure 8.17 and data from Figure A.2,

$$F_{3-56} = \frac{1}{2(2 \times 5)}[(5 \times 6)(0.51) - (4 \times 5)(0.45) + (2 \times 5)(0.32)] = 0.48$$

*When using small F_{1-2} charts, the accuracy of the calculated result is not the greatest. In this example, $F_{1-2} = 0.03$ means that $F_{1-2} \approx 0.03 \pm 0.01$. Large charts would improve the accuracy.

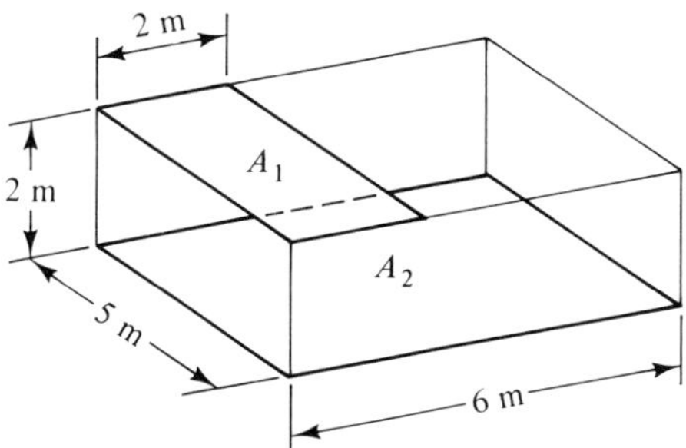

Figure 8.17

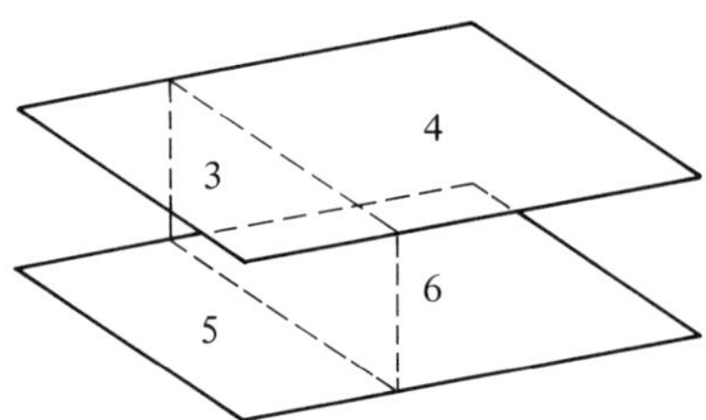

Figure 8.18

The preceeding type of calculations can be facilitated by use of geometric flux algebra, which is introduced in the next section.

8.11 Geometric Flux Algebra

The geometric flux is defined as

$$H_{ij} = A_i F_{ij} \tag{8.35}$$

where i is the emitter and j is the receiver. The rules for geometric flux algebra are easily developed from the shape factor relationships of Section 8.9.

$$H_{ij} = H_{ji}, \qquad \text{from equation (8.30)} \tag{8.36}$$

$$\sum_{j=i}^{n} H_{ij} = A_i, \qquad \text{from equation (8.31)} \tag{8.37}$$

In addition,

$$H_{1-23} = H_{1-2} + H_{1-3}, \qquad \text{from equation (8.32)} \tag{8.38}$$

and

$$H_{12-3} = H_{1-3} + H_{2-3}, \qquad \text{from equation (8.33)} \tag{8.39}$$

To check the validity of equation (8.39), simply expand it using the definition of geometric flux (8.35).

$$A_{12}F_{12-3} = A_1F_{1-3} + A_2F_{2-3}$$

which is equation (8.33). Finally, by repeated use of the preceding rules,

$$\begin{aligned} H_{12-34} &= H_{1-34} + H_{2-34} \\ &= H_{1-3} + H_{1-4} + H_{2-3} + H_{2-4} \end{aligned} \tag{8.40}$$

Example 8.11.1

Repeat Example 8.10.2 using geometric flux algebra (i.e., find H_{3-56}).

$$\begin{aligned} H_{3-56} &= H_{34-56} - H_{4-56} \\ &= H_{34-56} - H_{4-6} - H_{4-5} \\ &= H_{34-56} - H_{4-6} - H_{5-4} \\ &= H_{34-56} - H_{4-6} - (H_{5-43} - H_{5-3}) \end{aligned}$$

Now, by symmetry in Figure 8.18, $H_{5-43} = H_{3-56}$ and

$$2H_{3-56} = H_{34-56} - H_{4-6} + H_{5-3}$$

8.12 Special Reciprocity Relation

A glance back at equations (8.28) and (8.29) provides a reminder that the integrals for A_1F_{1-2} and A_2F_{2-1} are equivalent. Thus we have the basic reciprocity relation $A_1F_{1-2} = A_2F_{2-1}$ or $H_{1-2} = H_{2-1}$.

There is a special reciprocity relation, which arises again because of equality of integrals, that is necessary when dealing with complicated shapes. Referring to Figure 8.19, the special reciprocity relation states that

$$A_1F_{1-4} = A_2F_{2-3} \tag{8.41}$$

or

$$H_{1-4} = H_{2-3} \tag{8.42}$$

Consider radiation from dA_1 to dA_4:

$$\begin{aligned} dR_{dA_1-dA_4} &= \frac{I_b\, dA_1 \cos\phi_1\, dA_2 \cos\phi_2}{r^2} \\ &= \frac{I_b\,(dx_1\, dy) \cos\phi_1\, (dx_4\, dz) \cos\phi_2}{r^2} \end{aligned}$$

Integrate to get the total radiation from A_1 to A_4:

$$R_{A_1-A_4} = I_b \int_0^b \int_c^d \int_0^a \int_{x_1=0}^c \frac{\cos\phi_1 \cos\phi_2}{r^2}\, dx_1\, dy\, dx_4\, dz$$

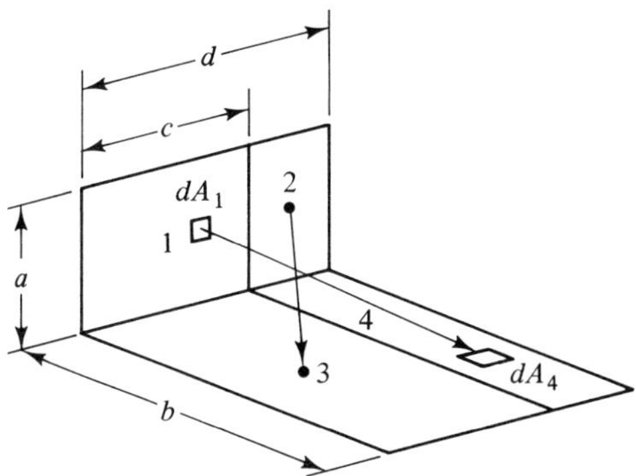

Figure 8.19

The shape factor is the ratio of $R_{A_1-A_2}$ to $\pi I_b A_1$:

$$F_{A_1-A_4} = \frac{R_{A_1-A_4}}{\pi I_b A_1} = \frac{1}{\pi A_1} \int_0^b \int_c^d \int_0^a \int_0^c \frac{\cos \phi_1 \cos \phi_2}{r^2}\, dx_1\, dy\, dx_4\, dz \tag{8.43}$$

Similarly, the shape factor from A_2 to A_3 is

$$F_{A_2-A_3} = \frac{1}{\pi A_2} \int_0^b \int_0^c \int_0^a \int_{x_2=c}^d \frac{\cos \phi_1 \cos \phi_2}{r^2}\, dx_2\, dy\, dx_3\, dz \tag{8.44}$$

Although the order of integration is altered, the integrals are the same. Thus $A_1F_{1-4} = A_2F_{2-3}$ or $H_{1-4} = H_{2-3}$.

The special reciprocity relation is valid for sets of areas that have the same limits of integration. In general, these subareas are formed by the insertion of planes perpendicular to the principal axes. Figures 8.20 and 8.21 demonstrate some possibilities. The limits of integration are shown with the figures, to demonstrate that each set has the same limits. In Figure 8.21, there are many other complementary sets for which the special reciprocity relation holds.

Figure 8.20

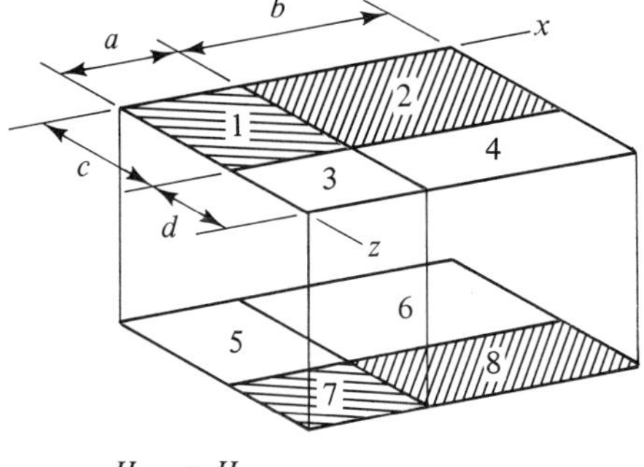

Limits of integration

$A_1 \to A_8$	x_1	from 0 to a
	z_1	0 to c
	x_8	a to $a + b$
	z_8	c to $c + d$
$A_2 \to A_7$	x_2	from a to $a + b$
	z_2	0 to c
	x_7	0 to a
	z_7	c to $c + d$

Limits of integration

$A_1 \rightarrow A_4$	x_1 from	0 to c
	y_1	a to $a+b$
	x_4	$c+d$ to $c+d+e$
	z_4	f to $f+g$
$A_2 \rightarrow A_3$	x_2 from	$c+d$ to $c+d+e$
	y_2	a to $a+b$
	x_3	0 to c
	z_3	f to $f+g$

$H_{1-4} = H_{2-3}$

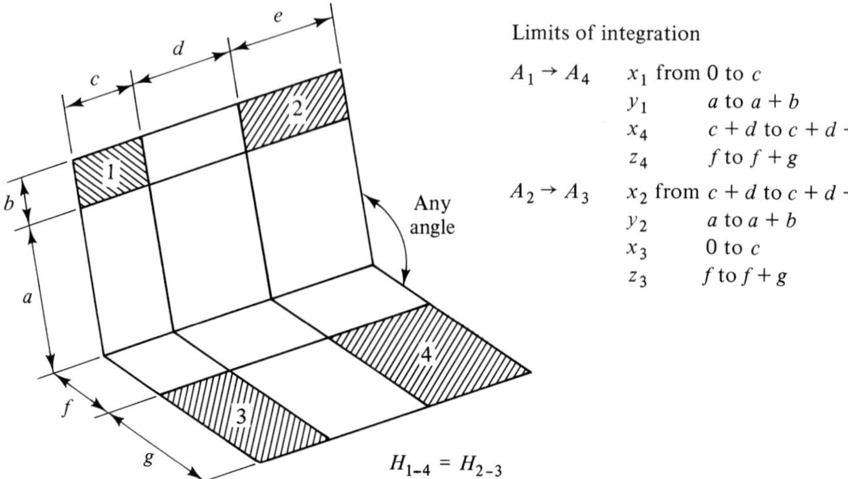

Figure 8.21

Example 8.12.1

Find the shape factor for A_1 radiating to A_{56} (Figure 8.22) in terms of areas and shape factors that can be found in Figure A.3.

$$\overset{\surd}{H_{1234-56}} = H_{1-56} + H_{2-56} + \overset{\surd}{H_{34-56}} \tag{8.45}$$

The checkmark ($\surd$) indicates those sets for which shape factors can be read from Figure A.3.

Now H_{1-56} is what we are trying to find, so in equation (8.45) we must manipulate H_{2-56} as it is the only other unknown.

$$\begin{aligned} H_{2-56} &= H_{2-5} + H_{2-6} \\ &= H_{1-6} + H_{2-6} \qquad \text{by special reciprocity relation} \\ &= (H_{1-56} - H_{1-5}) + (\overset{\surd}{H_{24-6}} - \overset{\surd}{H_{4-6}}) \\ &= H_{1-56} - (\overset{\surd}{H_{13-5}} - \overset{\surd}{H_{3-5}}) + \overset{\surd}{H_{24-6}} - \overset{\surd}{H_{4-6}} \end{aligned} \tag{8.46}$$

Figure 8.22

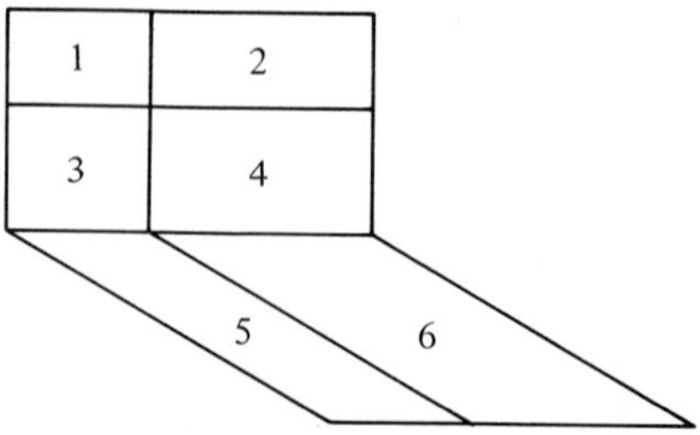

Substitute equation (8.46) into (8.45):

$$H_{1234-56} = H_{1-56} + (H_{1-56} - H_{13-5} + H_{3-5} + H_{24-6} - H_{4-6}) + H_{34-56}$$

Thus

$$2H_{1-56} = H_{1234-56} + H_{13-5} - H_{3-5} - H_{24-6} + H_{4-6} - H_{34-56}$$

and

$$F_{1-56} = \frac{1}{2A_1}(A_{1234}F_{1234-56} + A_{13}F_{13-5} - A_3F_{3-5} - A_{24}F_{24-6} + A_4F_{4-6} - A_{34}F_{34-56}) \tag{8.47}$$

Each one of these puzzles *is a puzzle*. No instructions are available directing a purely logical procedure from beginning to end. Anyone will make false starts or circle back such that one gets $H_{1-56} = H_{1-56}$. But practice does help!

8.13 Radiant Heat Exchange between Black Bodies

Because $\alpha = \varepsilon = 1$ for black bodies, and there is no reflection of radiation, calculations involving black-body radiation are relatively simple. Referring to Figure 8.23, energy radiated from A_1 to A_2 is

$$R_{A1-A2} = A_1F_{1-2}E_{b_1} = A_1F_{1-2}\sigma T_1^4 \tag{8.48}$$

Energy radiated from A_2 to A_1 is

$$R_{A2-A1} = A_2F_{2-1}E_{b_2} = A_2F_{2-1}\sigma T_2^4 \tag{8.49}$$

The net rate of energy transfer from A_1 to A_2 is

$$\begin{aligned} q_{1-2} &= R_{A1-A2} - R_{A2-A1} \\ &= \sigma(A_1F_{1-2}T_1^4 - A_2F_{2-1}T_2^4) \end{aligned} \tag{8.50}$$

By the basic reciprocity relation, $A_1F_{1-2} = A_2F_{2-1}$. Thus

$$q_{1-2} = \sigma A_1F_{1-2}(T_1^4 - T_2^4)$$

or (8.51)

$$q_{1-2} = \sigma A_2F_{2-1}(T_1^4 - T_2^4)$$

Figure 8.23 Radiation exchange between two black surfaces

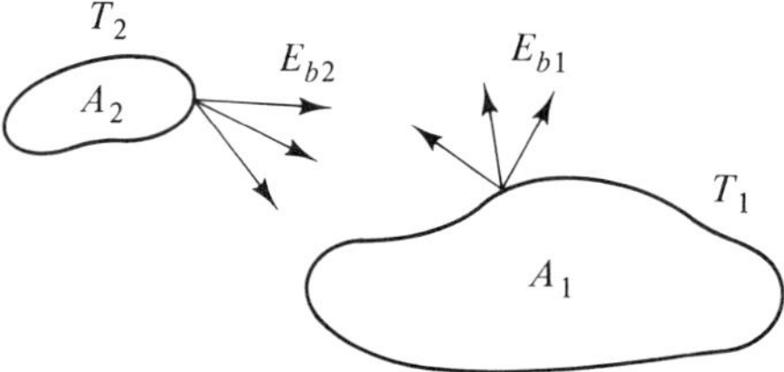

8.14 Real Surfaces and Gray Surfaces

We have considered black-body radiation and heat transfer between black bodies. In addition, we have noted that the emissivity for a real surface is always less than for a black body ($\varepsilon_{\text{real}} < 1$), and from Kirchhoff's law we have the statement that absorptivity equals emissivity for a real body in thermal equilibrium with its surroundings. However, bodies are not in thermal equilibrium when heat transfer is occurring.

The emissivity for a real body that is not in equilibrium with its surroundings is a function of its surface characteristics and its temperature. The absorptivity depends on those factors that affect its emissivity and on the character of the incoming radiation. The character of the incoming radiation changes with its source temperature. Thus *the emissivity and absorptivity of a body are not generally equivalent* in nonequilibrium situations. Evaluation of radiant heat transfer between real bodies can be done properly only by considering the radiation at each wavelength and integrating over the spectrum. The complexities arising are much beyond the scope of an introductory course. An idealization that is generally used to simplify radiation heat-transfer calculations is the concept of a gray body.

The total emissive power of a black body can be represented by the area under the spectral distribution curve for black-body radiation (see Figure 8.24).

$$E_b = \int_0^\infty E_{b\lambda}\, d\lambda \tag{8.52}$$

The total emissive power of a real body is similarly represented by the area

Figure 8.24 Monochromatic emissive power of real, black, and gray bodies

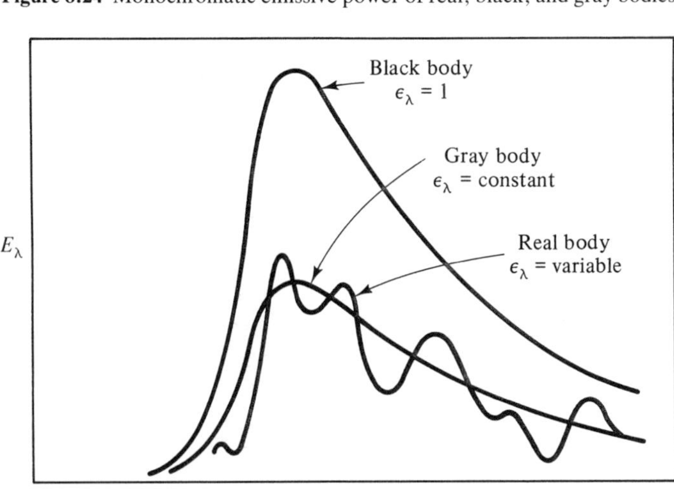

under the spectral distribution curve for the real-body emission:

$$E_{\text{real}} = \int_0^\infty \varepsilon_\lambda \, E_{b\lambda} \, d\lambda \tag{8.53}$$

where ε_λ is the monochromatic emissivity. In general, ε_λ is a variable, which means that real bodies radiate more or less efficiently depending on the wavelength of the radiation considered.

A *gray body* is defined as one for which the monochromatic emissivity is a constant, independent of wavelength.

$$\varepsilon_{\lambda\,\text{gray}} = \text{constant} \tag{8.54}$$

The gray-body concept is an approximation to real-body behavior, which simplifies the determination of total emissive power.

$$\begin{aligned} E_{\text{gray}} &= \int_0^\infty (\varepsilon_\lambda)_{\text{gray}} E_{b\lambda} \, d\lambda \\ &= (\varepsilon_\lambda)_{\text{gray}} \int_0^\infty E_{b\lambda} \, d\lambda \\ &= \varepsilon E_b \quad \text{or} \quad \varepsilon\sigma T^4 \end{aligned} \tag{8.55}$$

where the total emissivity $\varepsilon = (\varepsilon_\lambda)_{\text{gray}} = \text{constant}$. The value of $\varepsilon = (\varepsilon_\lambda)_{\text{gray}}$ would be chosen so that the area under the gray-body curve equals the area under the real-body curve. An additional important characteristic of the gray body, which results from the condition $\varepsilon_\lambda = \text{constant}$, is that the emissivity is equal to the absorptivity whether the radiating bodies are in thermal equilibrium or not.

$$\varepsilon_{\text{gray}} = \alpha_{\text{gray}} \tag{8.56}$$

8.15 Radiant Heat Exchange between Gray Bodies

Consider the exchange of thermal radiation between two infinite parallel gray planes (Figure 8.25). The radiant energy transferred per unit area from 1 and absorbed by 2 is

$$\frac{q'_{1-2}}{A} = \alpha_2 E_1 (1 + \rho_1\rho_2 + \rho_1{}^2\rho_2{}^2 + \cdots \rho_1{}^n\rho_2{}^n) = \frac{\alpha_2 E_1}{1 - \rho_1\rho_2} \tag{8.57}$$

Similarly, energy transferred from 2 and absorbed by 1 is

$$\frac{q'_{2-1}}{A} = \frac{\alpha_1 E_2}{1 - \rho_1\rho_2} \tag{8.58}$$

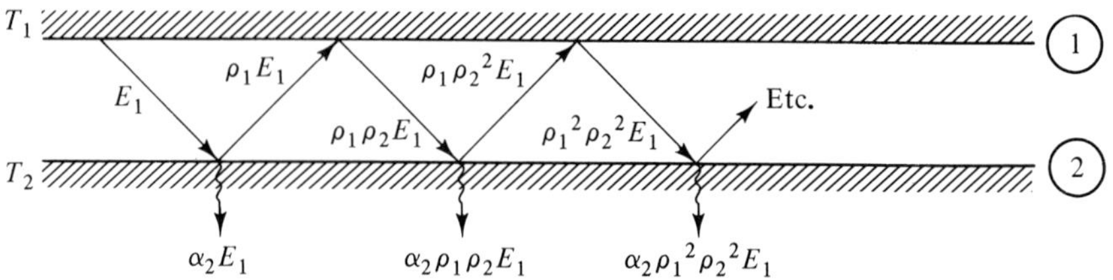

Figure 8.25 Radiation exchange between two gray planes

Thus, the net heat transfer from 1 to 2 is

$$\frac{q_{1-2}}{A} = \frac{q'_{1-2}}{A} - \frac{q'_{2-1}}{A} = \frac{\alpha_2 E_1 - \alpha_1 E_2}{1 - \rho_1 \rho_2} \tag{8.59}$$

Now $\rho_1 = 1 - \alpha_1 = 1 - \varepsilon_1$ and $\rho_2 = 1 - \varepsilon_2$. Thus

$$\begin{aligned}\frac{q_{1-2}}{A} &= \frac{\varepsilon_2 E_1 - \varepsilon_1 E_2}{1 - (1 - \varepsilon_1)(1 - \varepsilon_2)} \\ &= \frac{\varepsilon_2(\varepsilon_1 \sigma T_1^4) - \varepsilon_1(\varepsilon_2 \sigma T_2^4)}{1 - (1 - \varepsilon_1 - \varepsilon_2 + \varepsilon_1 \varepsilon_2)} \\ &= \frac{\sigma(T_1^4 - T_2^4)}{(1/\varepsilon_2) + (1/\varepsilon_1) - 1}\end{aligned} \tag{8.60}$$

While this method explains the process in clear physical fashion, it would be impossibly cumbersome for complex problems. The concept of radiosity and the use of electrical analogs make the handling of complex problems relatively simple.

8.16 Electrical Analog for Radiating Systems

Referring to Figure 8.26, let

G = irradiation = total radiant energy approaching a surface, W/m^2

J = radiosity = total radiant energy leaving a surface, W/m^2

Figure 8.26

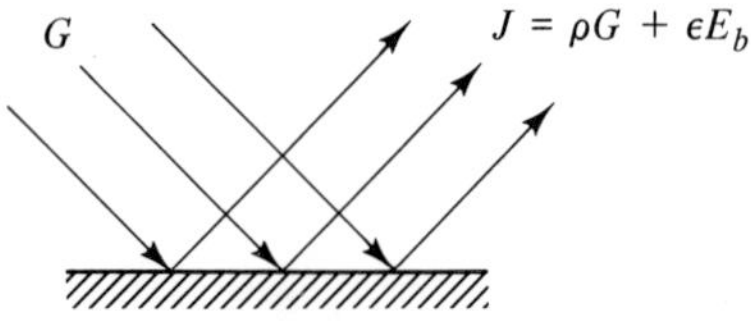

Referring to Figure 8.26, the radiosity J includes the portion of incoming energy that is reflected ρG and the radiation emitted from the surface due to its own temperature.

$$J = \rho G + \varepsilon E_b = (1 - \alpha)G + \varepsilon E_b = (1 - \varepsilon)G + \varepsilon E_b \tag{8.61}$$

The net energy flux leaving the surface is

$$\frac{q}{A} = J - G = (1 - \varepsilon)G + \varepsilon E_b - G = \varepsilon E_b - \varepsilon G \tag{8.62}$$

Solve for G in equation (8.61):

$$\begin{aligned} (1 - \varepsilon)G &= J - \varepsilon E_b \\ G &= \frac{J - \varepsilon E_b}{1 - \varepsilon} \end{aligned} \tag{8.63}$$

Substitute equation (8.63) into (8.62) to eliminate G:

$$\begin{aligned} \frac{q}{A} &= J - G = J - \frac{J - \varepsilon E_b}{1 - \varepsilon} \\ &= \frac{(J - \varepsilon J) - (J - \varepsilon E_b)}{1 - \varepsilon} = \frac{\varepsilon(E_b - J)}{1 - \varepsilon} \end{aligned} \tag{8.64}$$

Finally, the net rate of flow of energy from the surface can be expressed as

$$q_{\text{net}} = \frac{E_b - J}{(1 - \varepsilon)/A\varepsilon} \tag{8.65}$$

Comparing equation (8.65) with the equation for current flow through an electrical resistor, $i = \Delta V/R$, we arrive at an electrical analog where we can think of $E_b - J$ as the driving potential difference, $(1 - \varepsilon)/A\varepsilon$ as a surface resistance, and q as the resulting current (Figure 8.27).

Only a portion of the net radiant energy leaving surface i actually strikes surface j because of the shape factor. Thus

$$q'_{i-j} = A_i F_{i-j} J_i$$

Similarly, only a portion of the energy leaving j strikes i.

$$q'_{j-i} = A_j F_{j-i} J_j$$

Thus the net energy transfer from i to j is

$$q_{i-j(\text{net})} = q'_{i-j} - q'_{j-i} = A_i F_{i-j} J_i - A_j F_{j-i} J_j \tag{8.66}$$

Figure 8.27 Electrical analog for surface radiation

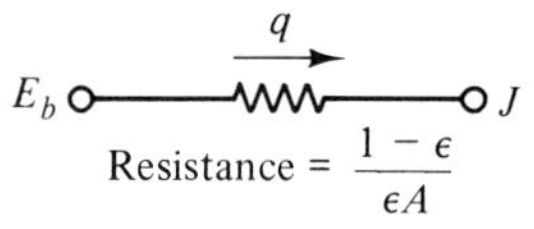

q

E_b — J

Resistance $= \dfrac{1 - \epsilon}{\epsilon A}$

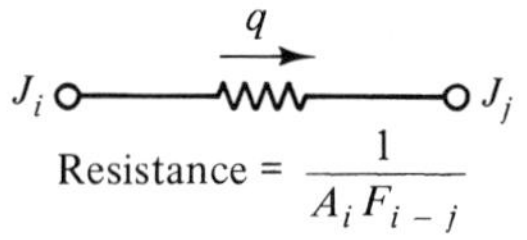

q

J_i —wwww— J_j

$$\text{Resistance} = \frac{1}{A_i F_{i-j}}$$

Figure 8.28 Electrical analog for radiation exchanged by two surfaces

Now, by the reciprocity relation, $A_i F_{i-j} = A_j F_{j-i}$. Thus

$$q_{i-j(\text{net})} = A_i F_{i-j}(J_i - J_j) = \frac{J_i - J_j}{1/(A_i F_{i-j})} \tag{8.67}$$

The electrical analog for equation (8.67) is shown in Figure 8.28, where $J_i - J_j$ is the driving potential and $1/(A_i F_{i-j})$ is considered a spatial resistance.

For radiant heat exchange between several gray surfaces, an analogous electrical network can be constructed made up of elements similar to the elements in Figures 8.27 and 8.28. Rates of heat transfer are then determined from solution of the electrical network.

Example 8.16.1

Solve for the heat transfer between two infinite parallel gray surfaces using the electrical analogy (see Figures 8.29 and 8.30). Considering the current q through the series of resistors,

$$q = \frac{E_{b1} - E_{b2}}{\Sigma \text{ resistances}} = \frac{E_{b1} - E_{b2}}{\dfrac{1 - \varepsilon_1}{\varepsilon_1 A_1} + \dfrac{1}{A_1 F_{1-2}} + \dfrac{1 - \varepsilon_2}{\varepsilon_2 A_2}} \tag{8.68}$$

For infinite parallel surfaces, the edge losses can be ignored, so the shape factor is unity ($F_{1-2} = 1$). Also, $A_1 = A_2 = A$. Therefore, equation (8.68) can be rewritten as

$$q = \frac{A(E_{b1} - E_{b2})}{\dfrac{1}{\varepsilon_1} - 1 + 1 + \dfrac{1}{\varepsilon_2} - 1} = \frac{A(E_{b1} - E_{b2})}{\dfrac{1}{\varepsilon_1} + \dfrac{1}{\varepsilon_2} - 1} \tag{8.69}$$

Finally, the heat transfer per unit area from surface 1 to surface 2 can be expressed as

$$\frac{q}{A} = \frac{E_{b1} - E_{b2}}{\dfrac{1}{\varepsilon_1} + \dfrac{1}{\varepsilon_2} - 1} = \frac{\sigma(T_1^4 - T_2^4)}{\dfrac{1}{\varepsilon_1} + \dfrac{1}{\varepsilon_2} - 1} \tag{8.70}$$

Note that the resulting equation (8.70) is the same as (8.60).

Figure 8.29 Parallel gray planes

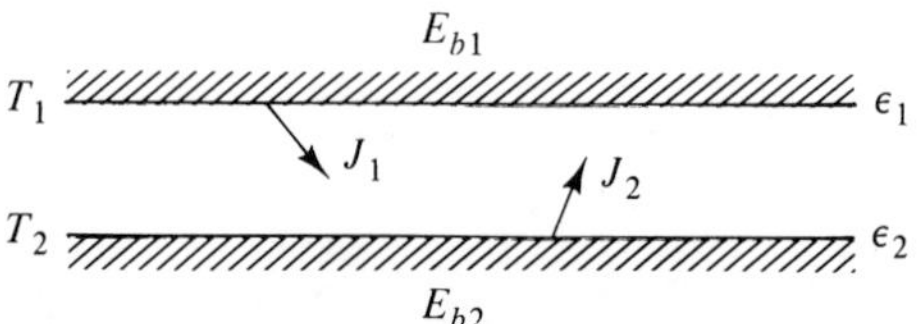

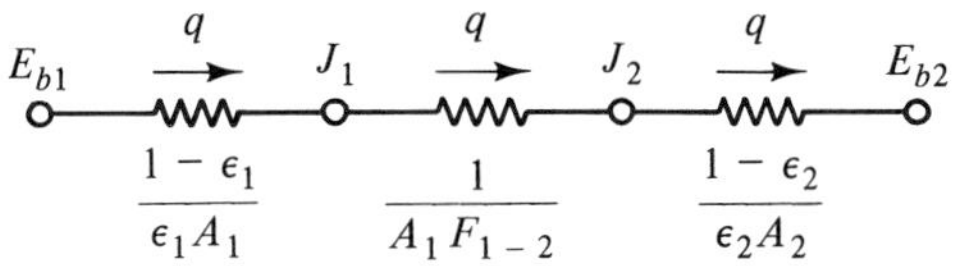

Figure 8.30 Electrical analog for parallel gray planes

Example 8.16.2

Determine the rate of heat transfer required (excluding the radiant energy transfer itself) to or from plates 1 and 2 in Figure 8.31 to maintain the temperatures at steady state.

Hints for constructing electrical analog network:

1. There is a node J for every radiating or receiving surface.
2. Each node is connected to every other node by a spatial resistance.
3. Each node is also connected to its E_b through a surface resistance.

To continue with the solution, we must first determine the shape factors. From Figure A.2, $F_{1-2} = 0.36$. Also, $F_{1-2} + F_{1-3} = 1$. Thus, $F_{1-3} = 1 - 0.36 = 0.64 = F_{2-3}$ (by geometric similarity). Kirchhoff's current law states that the sum of the currents entering a node is zero. From Figure 8.32,

$$q_1 = q_{1-2} + q_{1-3} \tag{8.71}$$

and

$$q_{1-2} = q_2 + q_{2-3} \tag{8.72}$$

From equation (8.71) and Figure 8.32,

$$\frac{E_{b_1} - J_1}{(1 - \varepsilon_1)/\varepsilon_1 A_1} = \frac{J_1 - J_2}{1/(A_1 F_{1-2})} + \frac{J_1 - J_3}{1/(A_1 F_{1-3})} \tag{8.73}$$

Similarly,

$$\frac{J_1 - J_2}{1/(A_1 F_{1-2})} = \frac{J_2 - E_{b_2}}{(1 - \varepsilon_1)/\varepsilon_2 A_2} + \frac{J_2 - J_3}{1/(A_2 F_{2-3})} \tag{8.74}$$

From equations (8.73) and (8.74), we can solve for the two unknowns J_1 and J_2. Note that $E_{b_1} = \sigma T_1^4$, $E_{b_2} = \sigma T_2^4$, $J_3 = 0$, and we have all the numerical data to

Figure 8.31 Two parallel, finite gray planes

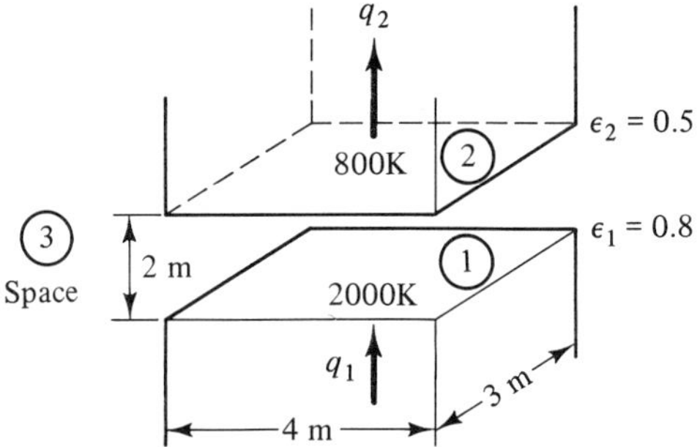

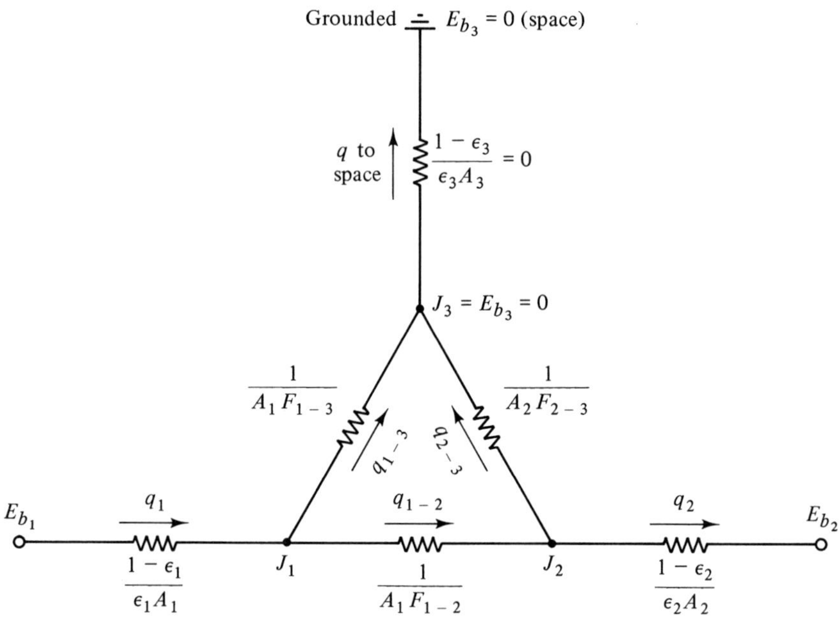

Figure 8.32 Electrical analog for two parallel, finite gray planes

solve for the resistances. Express J_2 as a function of J_1 from equation (8.73) and substitute into (8.74), and solve for J_1. Then solve for q_1 and q_2. Equation (8.73) with numerical values inserted gives

$$\frac{907\,040 - J_1}{0.02083} = \frac{J_1 - J_2}{0.2315} + \frac{J_1 - 0}{0.1302} \tag{8.75}$$

$$\frac{J_2}{0.2315} = \frac{J_1}{0.2315} + \frac{J_1}{0.1302} - \frac{907\,040 - J_1}{0.02083}$$

$$J_2 = 13.89J_1 - 10\,080\,641 \tag{8.76}$$

From equation (8.74),

$$\frac{J_1 - J_2}{0.2315} = \frac{J_2 - 23\,220}{0.08333} + \frac{J_2 - 0}{0.1302}$$

$$\frac{J_1}{0.2315} = \frac{J_2}{0.2315} + \frac{J_2}{0.08333} - \frac{23\,220}{0.08333} + \frac{J_2}{0.1302} \tag{8.77}$$

$$J_1 = 5.556J_2 - 64\,508 \tag{8.78}$$

Substitute equation (8.76) into (8.78):

$$J_1 = 5.556(13.89J_1 - 10\,080\,641) - 64\,508$$

$$= 77.17J_1 - 56\,072\,549$$

Thus $J_1 = 736\,150$ and from equation (8.76) $J_2 = 144\,483$. Finally,

$$q_1 = \frac{E_{b_1} - J_1}{(1 - \varepsilon_1)/\varepsilon_1 A_1} = \frac{907\,040 - 736\,150}{0.02083} = 8.20 \times 10^6 \text{ W}$$

and

$$q_2 = \frac{J_2 - E_{b2}}{(1 - \varepsilon_2)/\varepsilon_2 A_2} = \frac{144\ 482 - 23}{0.08333} = 1.455 \times 10^6 \text{ W}$$

Comments:

1. The results indicate that plate 1 must be heated and plate 2 must be cooled.
2. The answers can be no more accurate than the information supplied. Assuming that the data supplied are accurate to two significant figures, the answers should be rounded to $q_1 = 8.2 \times 10^6$ and $q_2 = 1.5 \times 10^6$ W.

8.16.1 Reradiating Surfaces. A *reradiating surface* is one that reflects and emits radiation at the same rate as it receives radiation and thereby experiences no heat gain or loss. An example might be the refractory walls in a furnace, where the conduction heat loss through the walls is negligible compared to the incident radiation. Consider the same two radiating surfaces as in Example 8.16.2, Figure 8.31, but now enclosed by four brick walls. No heat is supplied to or extracted from the brick walls. Surface 3 includes all four side walls (Figure 8.33).

Figure 8.33 Surfaces enclosed by reradiating walls

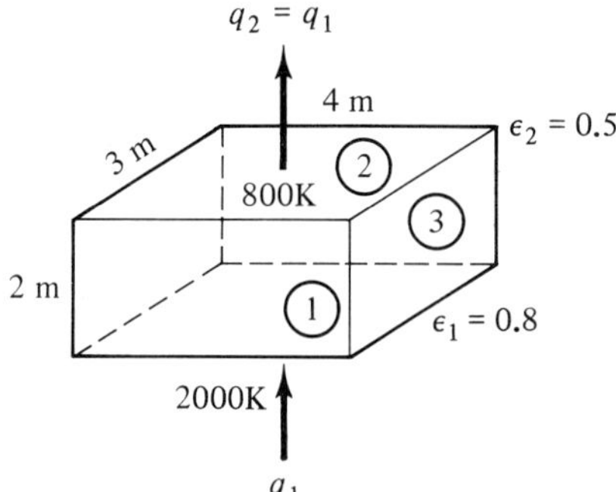

Because there is no net current flow out of node J_3 the solution (Figure 8.34) is more convenient than for the case involving radiation to space. For clarity, the analog circuit can be redrawn as in Figure 8.35.

The total resistance to radiant heat transfer from E_{b_1} to E_{b_2} is

$$R_{\text{total}} = \frac{1 - \varepsilon_1}{\varepsilon_1 A_1} + \frac{1}{A_1 F_{1-2} + 1\Big/\left(\dfrac{1}{A_1 F_{1-3}} + \dfrac{1}{A_3 F_{3-2}}\right)} + \frac{1 - \varepsilon_2}{\varepsilon_2 A_2} \tag{8.79}$$

In Example 8.16.2, we found $F_{1-2} = 0.36$ and $F_{1-3} = F_{2-3} = 0.64$. Also, by reciprocity, $A_3 F_{3-2} = A_2 F_{2-3}$. Thus

$$R_{\text{total}} = \frac{1 - 0.8}{0.8(12)} + \frac{1}{12(0.36) + 1\Big/\left(\dfrac{1}{12(0.64)} + \dfrac{1}{12(0.64)}\right)} + \frac{1 - 0.5}{0.5(12)}$$

$$= 0.0208 + 0.1225 + 0.0833 = 0.2266 \text{ m}^{-2}$$

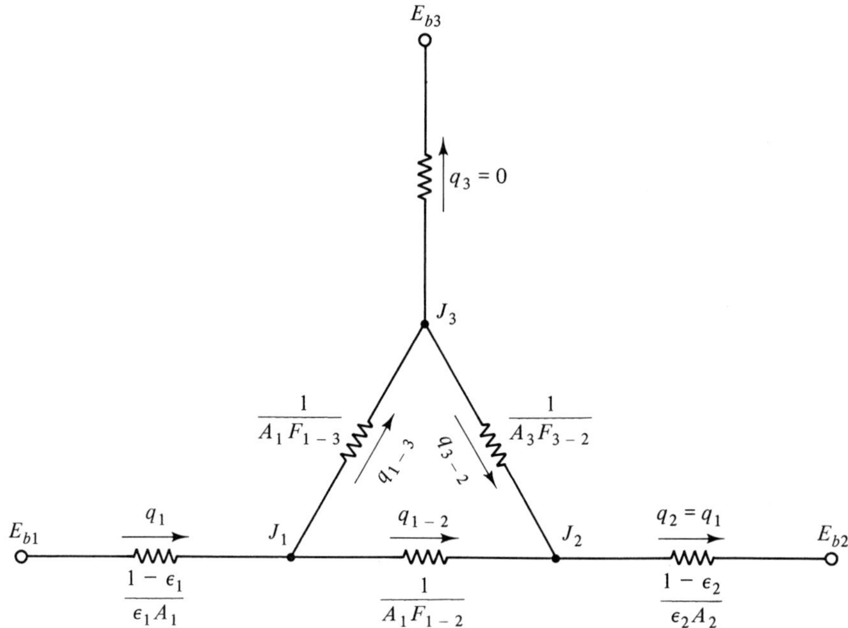

Figure 8.34 Electrical analog for two surfaces enclosed by reradiating walls

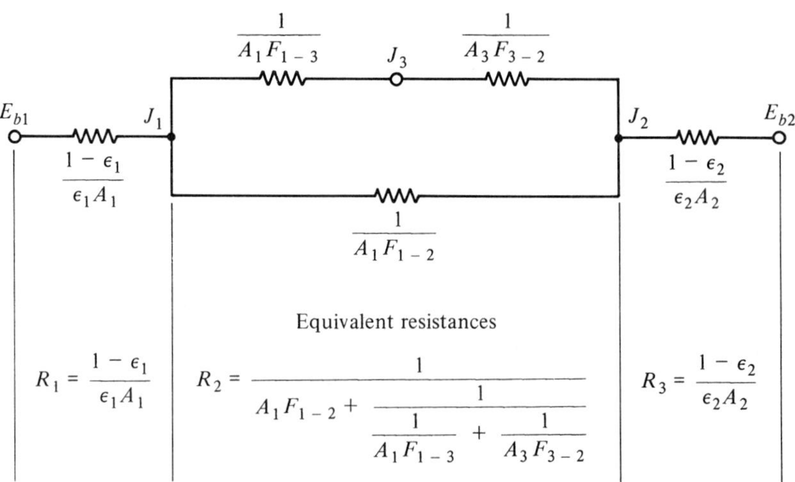

Figure 8.35 Analog of Figure 8.34 redrawn

The rate of heat transfer from surface 1 to surface 2 is

$$q_1 = q_2 = \frac{E_{b_1} - E_{b_2}}{R_{\text{total}}} = \frac{907\ 040 - 23\ 220}{0.2266} = 3\ 900\ 353 \text{ W}$$

This should be rounded off to

$$q_1 = q_2 = 3.9 \times 10^6 \text{ W}$$

Comparing this answer (where 3 is a reradiating surface) with the answer to Example 8.16.2 (where radiation to 3 escaped to space), we find that the energy required to keep plate 1 at 2000 K is now less. This is primarily because the diffuse radiation from the reradiating surface is partly returned to surface 1 rather than escaping to space.

8.16.2 Radiation Shielding.

Insertion of a thin layer of material between a hot radiator and its cooler receiver always reduces the amount of radiant heat transfer. If the shield is shiny (with low emissivity), it is even more effective in reducing the radiation transfer. Consider the heat transfer between large parallel plates with a shield between (Figure 8.36).

Referring to Example 8.16.1, equation (8.70), we have for heat transfer from 1 to 2 *with no shield*,

$$\frac{q_{1-2}}{A} = \frac{\sigma(T_1^4 - T_2^4)}{(1/\varepsilon_1) + (1/\varepsilon_2) - 1} \tag{8.70}$$

With the shield in place, we have

$$q_{1-2} = q_{1-3} = q_{4-2} \tag{8.80}$$

Now

$$\frac{q_{1-3}}{A} = \frac{\sigma(T_1^4 - T_3^4)}{(1/\varepsilon_1) + (1/\varepsilon_3) - 1}$$

and

$$\frac{q_{4-2}}{A} = \frac{\sigma(T_4^4 - T_2^4)}{(1/\varepsilon_4) + (1/\varepsilon_2) - 1}$$

Figure 8.36 Two planes with shield

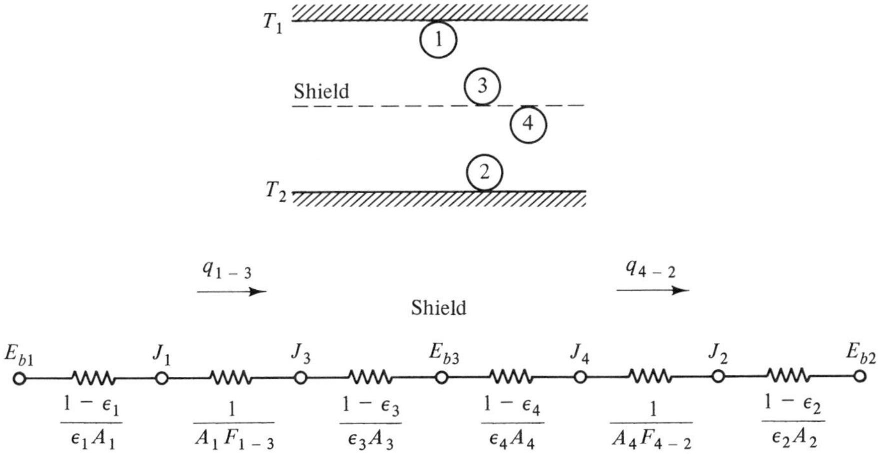

If emissivities $\varepsilon_1 = \varepsilon_2$ and $\varepsilon_3 = \varepsilon_4$, and $T_3 = T_4$, we have

$$T_{3,4}^4 = \frac{T_1^4 + T_2^4}{2} \tag{8.81}$$

and the heat flux is

$$\frac{q}{A} = \frac{\frac{1}{2}\sigma(T_1^4 - T_2^4)}{(1/\varepsilon_1) + (1/\varepsilon_3) - 1} \tag{8.82}$$

If all emissivities are equal, the radiant heat flux with shield is half the heat flux without a shield.

One of the most common applications of radiation shielding is in temperature measurement. Consider the case of hot air flowing through a pipe. The pipe wall is colder than the hot flowing air because the outside of the pipe is exposed to cool surroundings. When we place a thermometer bulb or thermocouple in the hot stream to measure the hot air temperature, heat will be radiated from the hot sensor to the cool pipe wall so that the sensor temperature will be less than the temperature of the hot air. At the same time, the sensor will be heated by convection heat transfer from the hot gas. The sensor will reach an equilibrium temperature that is lower than the hot air temperature, such that

$$\begin{aligned} q_{\text{convection}} &= q_{\text{radiation}} \\ hA(T_{\text{hot air}} - T_{\text{sensor}}) &= \varepsilon_{\text{sensor}} A\sigma(T_{\text{sensor}}^4 - T_{\text{pipe}}^4) \end{aligned} \tag{8.83}$$

The error in the temperature measured ($T_{\text{hot air}} - T_{\text{sensor}}$) can be reduced by placing a radiation shield around the sensor.

REFERENCES

1. Perry, R. H., and C. H. Chilton, *Chemical Engineer's Handbook*, 5th ed. New York: McGraw-Hill Book Company, 1973.
2. Parrish, A., *Mechanical Engineer's Reference Book*, 11th ed. London: Butterworth and Co. Ltd., 1973.

PROBLEMS

8.1. A black body has an emissive power of 12,000 W/m^2. What is the temperature of the body? *Answer:* 678 K

8.2. A Pyrex glass plate ($\varepsilon = 0.9$), 25 cm × 25 cm × 0.6 cm thick, is held at a temperature of 550°C. What is the emissive power of the glass at this temperature? *Answer:* 2.3×10^4 W/m^2

8.3. A beam of radiation equivalent to 2000 W/m^2 is incident on an object. The amount of energy absorbed and transmitted is 525 and 150 W/m^2, respectively. What is the reflectivity ρ of the object? *Answer:* 0.66

8.4. Ninety percent of a beam of thermal radiation is transmitted by fused quartz in the wavelength range from 0.2 to 4 μm. If a source of black-body radiation is viewed through a quartz window, what radiant heat flux is transmitted through the window if the source temperature is (a) 900°C, (b) 400°C? *Answer:* (a) 5.7×10^4 W/m²; (b) 2.1×10^3 W/m²

8.5. A gray plate ($\varepsilon = 0.6$) is attached to an upper surface of a high-flying jet airplane so that the plate is insulated from the fuselage. If the radiant heat flux from the sun at that point in space is 1400 W/m², determine the steady-state temperature of the plate. Assume outer space is equivalent to a black body at a temperature of zero degrees absolute. *Answer:* 396 K

8.6. A black body is held at a temperature of 500°C. Determine:
(a) Wavelength at which the maximum monochromatic emissive power occurs
(b) Emissive power at that wavelength
(c) Total emissive power of the black body
(d) Fraction of the total radiant emission that occurs between the wavelength 1.5 and 3.5 μm. *Answer:* (a) 3.86 μm; (b) 3546 W/m²-μm; (c) 2.02×10^4 W/m²; (d) 0.205

8.7. **(a)** What is the net radiant exchange between two large parallel black planes, one at 120°C and the other at 500°C? *Answer:* 1.89×10^4 W/m²
(b) If a third black plane is inserted between the two planes of part a, what is the ratio of the radiant energy transferred to that in part a? At steady conditions the intermediate plane takes up a temperature T_3. *Answer:* $\frac{1}{2}$

8.8. Find the shape factor F_{1-2}. *Answer:* 0.06

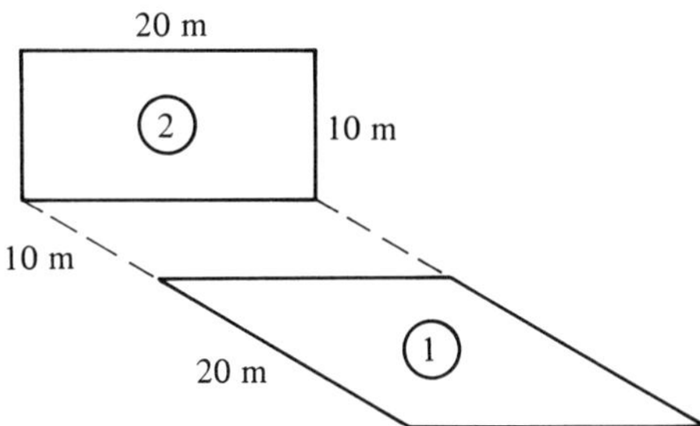

8.9. Find F_{1-2}. *Answer:* 0.06

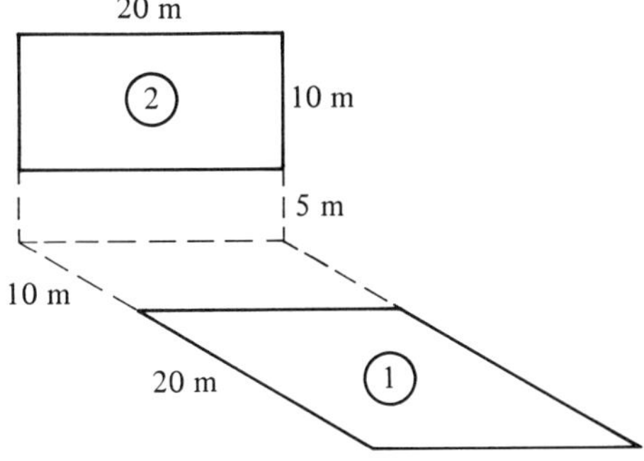

8.10. Find F_{1-2}. *Answer:* 0.020

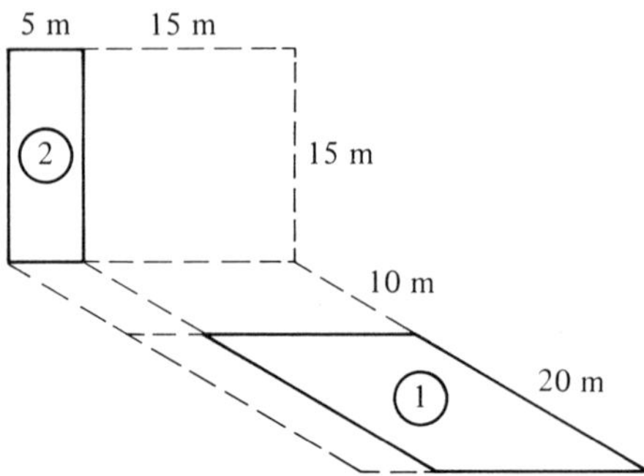

8.11. Find F_{1-2}. *Answer:* 0.062

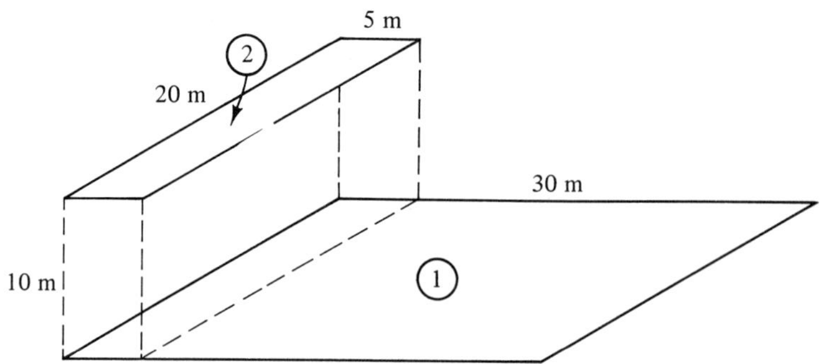

8.12. Find F_{1-2}. *Answer:* 0.022

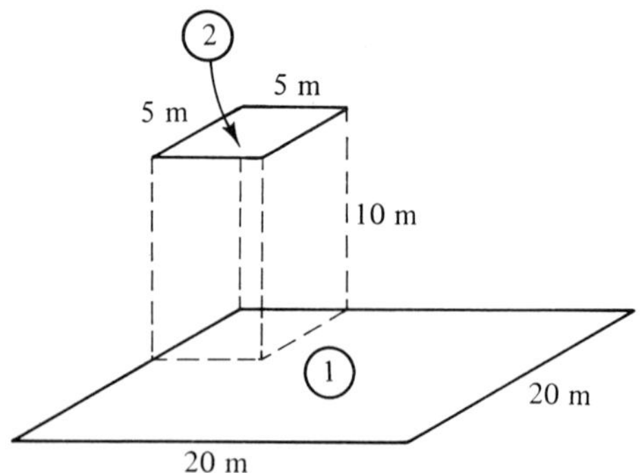

8.13. Two black rectangles, parallel and directly opposed, 5 m by 8 m and 4 m apart, exchange radiant energy. If the temperature of one rectangle is 200°C and the other is 400°C, at what rate is energy exchanged? What is the rate at which the 400°C rectangle is losing energy? Assume the surroundings are black at 25°C. *Answer:* 1.16×10^5, 4.16×10^5 W

8.14. A black rectangle, 5 m by 10 m, is normal to a second black rectangle that is 7 m by 10 m. The two rectangles have their 10-m side in common. If the

5 × 10 rectangle is at 300°C and the 7 × 10 rectangle is at 600°C, what is the rate of radiant energy exchange between them? At what rate must heat flow into the 7 × 10 rectangle to keep its temperature constant at 600°C? Assume black surroundings at 25°C. *Answer:* 3.57 × 10^5, 2.20 × 10^6 W

8.15. A cubical container, 5 m on a side, has all its surfaces black. One wall is kept at 20°C, the wall opposite at 35°C, and all the remaining walls are at 26°C. At what rate does the 35°C wall lose heat? *Answer:* 1.6 × 10^3 W

8.16. An enclosure consists of a cylinder, 30 cm in diameter by 60 cm long, closed at the ends by reradiating surfaces. The inside surface of the cylinder is black and at a constant temperature of 120°C. What is the steady-state temperature of the end surfaces? *Answer:* 120°C

8.17. A small industrial furnace is 3 m wide by 3 m long by 4 m high. The floor acts as a gray plane at 800°C (ε = 0.6), and the ceiling acts as a gray plane at 450°C (ε = 0.8). All other surfaces can be considered as a single reradiating wall. Determine the rate of radiant energy exchange between the floor and ceiling, as well as the steady-state temperature of the reradiating walls. *Answer:* 2.00 × 10^5 W, 648°C

8.18. Two rectangles, 4 m by 5 m, parallel and directly opposed, are spaced 5 m apart. One rectangle is held at 250°C and the other at 450°C. Find the net exchange of radiation between the two rectangles if they are:

(a) Black and unenclosed in a radiation-free environment.

(b) Black and enclosed by a single reradiating surface.

(c) Gray (ε = 0.6 for 250°C surface, ε = 0.8 for 450°C surface) and unenclosed in a radiation-free environment.

(d) Gray, as in part c, and enclosed by a single reradiating surface. *Answer:* (a) 3.82 × 10^4 W; (b) 1.32 × 10^5 W; (c) 3.10 × 10^4 W; (d) 8.57 × 10^4 W

8.19. Two 5-m by 5-m squares, directly opposed, are spaced 4 m apart. One square (ε = 0.6) is held at 220°C, while the other (ε = 0.8) is kept at 35°C. A third square of the same dimensions (ε = 0.9) is placed equidistant between the other two squares. Assume that the temperature is the same on both sides of the third square and that the surroundings are black at 25°C.

(a) What is the rate of radiant energy gain by the low-temperature square before the third square is inserted?

(b) What is the rate of energy gain by the low-temperature square after the third plane is in place?

(c) What is the steady-state temperature of the third square? *Answer:* (a) 8.20 × 10^3 W; (b) 3.15 × 10^3 W; (c) 85°C

8.20. A cylindrical cavity, 15 cm in diameter and 25 cm long, is closed at one end by a circular disk and at the other end by a circular disk containing a 7.5-cm-diameter concentric hole. The surroundings may be considered radiation free. Find the rate at which radiant energy leaves the hole if

(a) All interior surfaces are black at 300°C.

(b) The base of the cavity is gray (ε = 0.5) at 300°C and all other surfaces are reradiating.

(c) All interior surfaces are gray (ε = 0.5) at 300°C. *Answer:* (a) 27.0 W; (b) 18.5 W; (c) 26.2 W

8.21. A cylindrical cavity, 25 cm in diameter by 50 cm long, has the ends closed by circular disks. One end ($\varepsilon = 0.7$) is held at 100°C, the other end ($\varepsilon = 0.5$) at 200°C, and the cylindrical walls ($\varepsilon = 0.8$) are at a temperature of 400°C. Determine the rate at which energy is given up by each of the three surfaces. *Answer:* $q_1 = -342$ W, $q_2 = -197$ W, $q_3 = 540$ W

Steady-State Diffusion of Mass

9

The transfer of mass within a fluid mixture or across a phase boundary is a process that occurs in many systems of engineering interest. For instance, dispersion of gases into the atmosphere from the stack of a thermal power station and separation of the components of crude oil by rectification are just two of many examples that could be cited.

In a process like dispersion of gases from stacks, the mechanism of mass transfer involves both molecular diffusion and convection, with convection probably the predominant mechanism in this instance. However, in many cases convective forces are negligible, and molecular diffusion is the major mode of mass transfer.

The emphasis in this chapter is on the process of molecular diffusion. This process results from a concentration gradient and, as such, is analogous to heat transfer by conduction, which results from a temperature gradient. In fact, it was the similarity of these two processes recognized by A. Fick that led him to put diffusion on a quantitative basis by adopting the equation for heat conduction derived earlier by Fourier.

9.1 Fick's Law of Diffusion

In an isotropic medium, Fick's law is based on the hypothesis that the rate of transfer of the diffusing substance through a unit area is proportional to the concentration gradient measured normal to this area. We were introduced to this law in Chapter 1 as equation (1.10), where the flux was expressed in molar units. For a system of two components, A and B, this equation is written in molar units in the following way for diffusion in the y direction only.

$$J_A = -CD_{AB}\frac{\partial x_A}{\partial y} \tag{9.1}$$

where J_A = molar rate of diffusion of A per unit area in the y direction

C = total molar density of the mixture

D_{AB} = diffusion coefficient of A with respect to B

x_A = mole fraction of A

J_A is usually referred to as the *molar flux* of A. The product of C and x_A is the moles of A per unit volume or the molar concentration of A. The negative sign indicates that A diffuses spontaneously in the direction of a decrease in concentration.

9.2 Diffusion Coefficient

The proportionality factor D_{AB} in Fick's law is a property of the system whose value depends on the pressure, temperature, and composition of the system. Table A.6 lists some values for gases.

When experimental values are not available, correlations may be used to estimate values for D. The Gilliland equation, which follows, is one such correlation for diffusion in binary mixtures of gases (2, 3).

$$D_{AB} = \frac{1.57 \times 10^3\ T^{3/2}}{P(V_A{}^{1/3} + V_B{}^{1/3})^2}\left(\frac{1}{M_A} + \frac{1}{M_B}\right)^{1/2} \tag{9.2}$$

where D_{AB} = diffusion coefficient in cm^2/hr

T = temperature, K

P = absolute pressure, kPa

V = molecular volume (as given in Table 9.1)

M = molecular weight

The form of equation (9.2) was derived from the kinetic theory of gases, assuming that the distance between the centers of two molecules in collision is proportional to the sum of the cube roots of the volumes of the molecules in contact. The molecular volumes were determined as outlined by Arnold (1), whose report should be studied for details of the methods used. Essentially, these methods assume that V is the volume of 1 mole of liquid at its boiling point. For purposes of calculating V from knowledge of the

TABLE 9.1
ATOMIC VOLUMES
For use with equation (9.2)

Air (molecular volume)	29.9
Bromine	27.0
Carbon	14.8
Chlorine: terminal as in R—Cl	21.6
medial as in R—CHCl—R′	24.6
Fluorine	8.7
Hydrogen: in compounds	3.7
in hydrogen molecule	7.15
Nitrogen: in primary amines	10.5
in secondary amines	12.0
in nitrogen molecule	15.6
Oxygen: doubly bound in organic compounds	7.4
coupled to two other elements	
in methyl esters	9.1
in ethyl ethers	9.9
in higher esters and ethers	11.0
in acids	12.0
in union with S, P, N	8.3
in oxygen molecule	12.8
Sulfur	25.6
Water[a] (molecular volume)	18.9
Iodine	37.0

Note: Deduct for ring compounds:
Five-membered ring (furan), deduct 11.5
Six-membered ring (benzene), deduct 15.0
Naphthalene ring, deduct 30.0

[a] Taken from Arnold (1).

structure of the molecules involved, various rules and values for the atomic volumes were used. Table 9.1 summarizes some of these rules and atomic volumes.

Diffusion coefficients for water vapor in air are well correlated, for temperatures up to 1100°C, by an equation proposed by Spalding (3).

$$D_{wa} = \frac{33.2T^{2.5}}{P(T + 245)} \tag{9.3}$$

where the units are the same as in equation (9.2).

For organic molecules in liquid solution, a correlation due to Wilke (3) may be used in the absence of experimental data.

$$D_{AB} = \frac{1.7 \times 10^{-3}\, T}{\mu(V_A{}^{1/3} - k_1)} \tag{9.4}$$

where the units are as in equation (9.2).

μ = dynamic viscosity, kg/m-s

V_A = molecular volume of the solute obtained from Table 9.1

k_1 = 2.0, 2.46, and 2.84 for dilute solutions in water, methyl alcohol, and benzene, respectively

9.3 Stefan's Experiment

An early experiment used by Stefan (4) for measuring the rate of diffusion of a vapor through an inert gas is worth reviewing briefly. The experiment is illustrated in Figure 9.1. A layer of liquid such as water is placed at the closed bottom of a glass tube, while a flow of air containing water vapor at a constant concentration is passed over the open end of the tube. The pressure inside the tube is constant and equal to the outside pressure. The temperatures of the water, tube, and air are kept the same and constant and at a value so that the vapor pressure of the water is greater than the partial pressure of water vapor in the air stream. Under these conditions, the partial pressure of the water vapor decreases from the surface of the liquid water to the top of the tube. Since the sum of the partial pressure of the water vapor and the air is equal to the total pressure P, which is constant, the partial pressure of the air has its highest value at the top of the tube and its smallest value at the water surface. The two partial pressure gradients in the tube for steady conditions are as illustrated in Figure 9.1.

The molar flux of water vapor due to diffusion up the tube is given by equation (9.1).

$$J_w = -CD_{wa}\frac{\partial x_w}{\partial y} \tag{9.1}$$

where the subscripts w and a refer to water and air, respectively.

Figure 9.1 Stefan's experiment for diffusion of water vapor through air

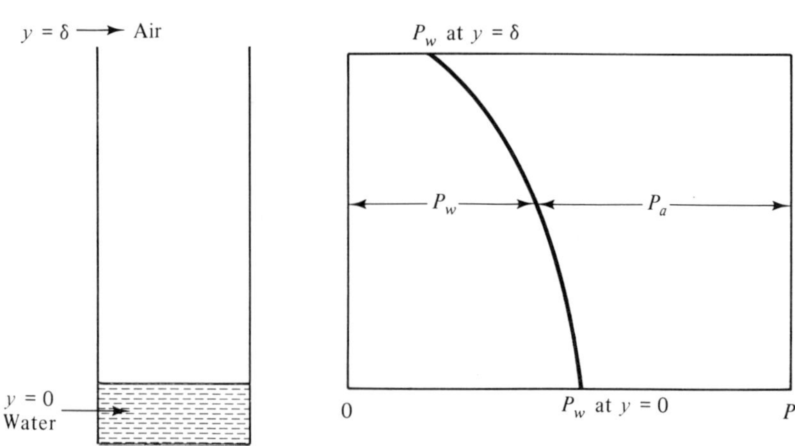

Equation (9.1) must also be valid for the air, which has a concentration gradient of opposite sign to that of the water vapor, indicating there must be a flow of air into the tube. However, no air can leave at the bottom since the tube is closed at that end. We deduce, therefore, under steady conditions that the diffusion of air down the tube must be balanced by an equal flow of air up the tube so that the net flow of air is zero. This latter flow cannot be diffusive since this requires a negative concentration gradient. It must therefore be a type of bulk or convective flow.

Suppose the velocity of this convective flow is U. The molar flux of water vapor transported by this convective flow is then UC_w, where C_w is the molar density of water vapor at some cross section in the tube. The total flux of water vapor up the tube, N_w, can now be expressed as the sum of that due to diffusion and to convection.

$$N_w = -CD_{wa}\frac{\partial x_w}{\partial y} + UC_w \tag{9.5}$$

The bulk-flow term in equation (9.5) may be rewritten in the following way. The total molar flux by bulk flow for the mixture is the sum of the flux for each component of the mixture.

$$N = N_w + N_a$$

N is also the product of the velocity of bulk flow and the total molar density of the mixture.

$$N = UC$$

The velocity of bulk flow can thus be expressed as

$$U = \frac{1}{C}(N_w + N_a) \tag{9.6}$$

If equation (9.6) is multiplied by the molar density of the water vapor, the following is obtained:

$$UC_w = \frac{C_w}{C}(N_w + N_a) = x_w(N_w + N_a) \tag{9.7}$$

which can then be used to rewrite the bulk-flow term in (9.5).

$$N_w = -CD_{wa}\frac{\partial x_w}{\partial y} + x_w(N_w + N_a) \tag{9.8}$$

Equation (9.8) is the differential equation that describes the transport of water vapor in the Stefan tube. When integrated for steady conditions, it gives a way to calculate the diffusion coefficient, D_{wa}, from a measure of the flux, N_w.

9.4 Steady-State Diffusion through a Stagnant Gas

In the Stefan tube, when steady conditions have been attained, the diffusion of air into the tube is balanced by the convective flow of air out of the tube. The air is therefore stagnant and

$$N_a = 0$$

Equation (9.8) thus becomes

$$N_w = -CD_{wa}\frac{\partial x_w}{\partial y} + x_w N_w$$

or, solving for N_w,

$$N_w = \frac{-CD_{wa}}{1 - x_w}\frac{\partial x_w}{\partial y} \tag{9.9}$$

For steady conditions, the flux of water vapor is the same at every cross section along the tube. This may be expressed in the following way:

$$\frac{\partial N_w}{\partial y} = 0$$

or

$$\frac{d}{dy}\left(\frac{-CD_{wa}}{1 - x_w}\frac{\partial x_w}{\partial y}\right) = 0 \tag{9.10}$$

For a gas mixture at constant temperature and pressure, the total molar density C is a constant. Assuming the diffusion coefficient D_{wa} also to be constant permits equation (9.10) to be written as

$$\frac{d}{dy}\left(\frac{1}{1 - x_w}\frac{\partial x_w}{\partial y}\right) = 0 \tag{9.11}$$

This is readily integrated to give

$$\frac{1}{1 - x_w}\frac{\partial x_w}{\partial y} = C_1 \tag{9.12}$$

Integration again results in

$$-\ln(1 - x_w) = C_1 y + C_2 \tag{9.13}$$

where C_1 and C_2 are constants of integration.

To evaluate C_1 and C_2, the following two boundary conditions may be used.

For $y = 0$, x_w = mole fraction of water vapor at the surface of the liquid water, x_{ws}

For $y = \delta$, x_w = mole fraction of water vapor at the top of the tube, $x_{w\delta}$

From the first boundary condition and equation (9.13), we obtain

$$\text{For } y = 0, \qquad -\ln(1 - x_{ws}) = C_2$$

and from the second boundary condition and (9.13),

$$\text{For } y = \delta, \qquad -\ln(1 - x_{w\delta}) = C_1\delta + C_2$$

Solving for C_1 and C_2 and substituting back into (9.13) gives the following general solution:

$$-\ln(1 - x_w) = \frac{y}{\delta}\ln\left(\frac{1 - x_{ws}}{1 - x_{w\delta}}\right) - \ln(1 - x_{ws})$$

This may be rewritten in either of the following ways:

$$\ln\left(\frac{1 - x_{ws}}{1 - x_w}\right) = \frac{y}{\delta}\ln\left(\frac{1 - x_{ws}}{1 - x_{w\delta}}\right)$$

or (9.14)

$$\frac{1 - x_w}{1 - x_{ws}} = \left(\frac{1 - x_{w\delta}}{1 - x_{ws}}\right)^{y/\delta}$$

Since $x_w + x_a = 1$, equation (9.14) may also be written as

$$\frac{x_a}{x_{as}} = \left(\frac{x_{a\delta}}{x_{as}}\right)^{y/\delta} \tag{9.15}$$

Expressions (9.14) and (9.15) give the composition of the gas along the tube when steady conditions are in effect.

The rate of evaporation of the water at the bottom of the tube can be obtained by integrating equation (9.9) directly for steady-state conditions. Thus

$$N_w\bigg|_{y=0} = N_{ws} = \frac{-CD_{wa}}{1 - x_w}\frac{\partial x_w}{\partial y} \tag{9.16}$$

Since N_{ws} is a constant, the integration is carried out as follows:

$$N_{ws}\int_0^\delta dy = -CD_{wa}\int_{x_{ws}}^{x_{w\delta}} \frac{dx_w}{1 - x_w}$$

Solving for N_{ws} gives

$$N_{ws} = \frac{CD_{wa}}{\delta}\ln\frac{x_{a\delta}}{x_{as}} \tag{9.17}$$

where the relation $x_a + x_w = 1$ has also been used.

Equation (9.17) may be written in various ways. One of these is to use the log-mean mole fraction of air across the cell:

$$(x_a)_{lm} = \frac{x_{a\delta} - x_{as}}{\ln(x_{a\delta}/x_{as})} \tag{9.18}$$

Upon substitution, equation (9.17) becomes

$$N_{ws} = \frac{CD_{wa}}{\delta} \frac{x_{a\delta} - x_{as}}{(x_a)_{lm}} \tag{9.19}$$

or, since $x_w + x_a = 1$,

$$N_{ws} = \frac{CD_{wa}}{\delta} \frac{x_{ws} - x_{w\delta}}{(x_a)_{lm}} \tag{9.20}$$

For certain kinds of problems, it is more convenient to express concentrations in terms of partial pressures rather than mole fractions. Thus, using the ideal gas law,

$$PV = n\,RT$$

where n = number of moles of gas in the volume V. The total molar density is

$$C = \frac{n}{V} = \frac{P}{RT}$$

Also, the mole fraction of a component of the gas mixture is given by the ratio of the partial pressures.

$$x_w = \frac{P_w}{P}$$

where P_w = partial pressure of water vapor in the gas mixture.
Using these relationships, equation (9.18) can be rewritten as

$$(x_a)_{lm} = \frac{P_{a\delta} - P_{as}}{P \ln(P_{a\delta}/P_{as})} = \frac{1}{P}(P_a)_{lm}$$

and equation (9.20) becomes

$$N_{ws} = \frac{PD_{wa}}{RT\delta(P_a)_{lm}}(P_{ws} - P_{w\delta}) \tag{9.21}$$

Equation (9.21) may be written in terms of a potential and a mass transfer resistance, which is analogous to equation (5.5) for conduction of heat through a wall.

$$N_{ws} = \frac{\Delta P_w}{R_m} \tag{9.22}$$

where the mass transfer resistance is given by

$$R_m = \frac{RT\delta(P_a)_{lm}}{PD_{wa}}$$

The discussion of Stefan's experiment and the expressions derived for calculating the flux have been in terms of water vapor diffusing through air.

Any two components, one of which is liquid at the temperature of the experiment, could have been used. In fact, in just this way the diffusion coefficients for a number of vapors in various gases have been determined.

Example 9.4.1

A Stefan tube is used to measure the rate of diffusion of benzene into air at 20°C and 101.3 kPa pressure. The distance between the benzene–air interface and the top of the tube is 15 cm. The vapor pressure of benzene at 20°C is 97 mm Hg. What is the expected flux of benzene in the tube? Equation (9.21) may be used for this calculation. In Table A.6, the diffusion coefficient for benzene in air is given as 278 cm²/hr at 0°C and 101.3 kPa. Use equation (9.2) to correct for temperature.

$$D_{ba}(20°\text{C}) = 278 \left(\frac{293}{273}\right)^{1.5} = 309 \text{ cm}^2/\text{hr}$$

The log-mean partial pressure of air is

$$(P_a)_{lm} = \frac{760 - (760 - 97)}{\ln[760/(760 - 97)]} = 710 \text{ mm Hg}$$

Benzene flux using equation (9.21) is

$$N_{ba} = \frac{101.3 \times 309 \times 10^{-4} \times 10^3}{8314 \times 293 \times (15/100) \times 710}(97 - 0) = 1.17 \times 10^{-3} \text{ kg moles/m}^2\text{-hr}$$

9.5 Equimolar Counterdiffusion

A number of interesting engineering problems involve mass transfer in which two components diffuse in opposite directions and in approximately equimolar amounts. For instance, combustion of a particle of coal occurs by means of the following chemical reaction:

$$C + O_2 \rightarrow CO_2$$

The process requires that, for every mole of oxygen diffusing toward the surface of the particle, 1 mole of carbon dioxide must diffuse away from that surface. A similar situation exists in a distillation or rectifying column where the mass transfer between vapor and liquid phases is approximately equimolar. Two solid metals in contact will also diffuse into each other at rates that are essentially equimolar.

Suppose we consider a system of two components, A and B, diffusing through a tube that is open at both ends. The flux of component A is given by equation (9.8):

$$N_A = -CD_{AB}\frac{\partial x_A}{\partial y} + x_A(N_A + N_B) \qquad (9.8)$$

For equimolar counterdiffusion,

$$N_A = -N_B$$

and there is no bulk-flow term. Thus

$$N_A = -CD_{AB}\frac{\partial x_A}{\partial y} \tag{9.23}$$

Integrating for constant temperature and pressure and for steady conditions,

$$\int_0^L dy = -\frac{CD_{AB}}{N_A}\int_{x_{A1}}^{x_{A2}} dx_A$$

which gives

$$N_A = \frac{CD_{AB}}{L}(x_{A1} - x_{A2}) \tag{9.24}$$

where L = length of the diffusion path.

If the diffusing mixture is a gas, equation (9.24) may be written in terms of partial pressures and the ideal gas law.

$$C = \frac{P}{RT}, \qquad x_A = \frac{P_A}{P}$$

where P = total pressure of the gaseous mixture

P_A = partial pressure of A in the mixture

Substituting into equation (9.24) and simplifying gives

$$N_A = \frac{D_{AB}}{RTL}(P_{A1} - P_{A2}) \tag{9.25}$$

We may also write this in terms of a potential and a resistance to mass transfer.

$$N_A = \frac{\Delta P_A}{R_m}$$

where R_m = mass transfer resistance = RTL/D_{AB}.

9.6 Diffusion into an Infinite Stagnant Medium

We will discuss three problems involving diffusion from a spherical particle into an infinite body of gas, which we will assume to be stagnant. The purpose in doing this is to demonstrate how to use the diffusion equation to derive the differential equations that describe these processes. The solutions obtained, however, will be largely academic because a large body of gas in which there are no convection currents is unlikely to be found in practice. In

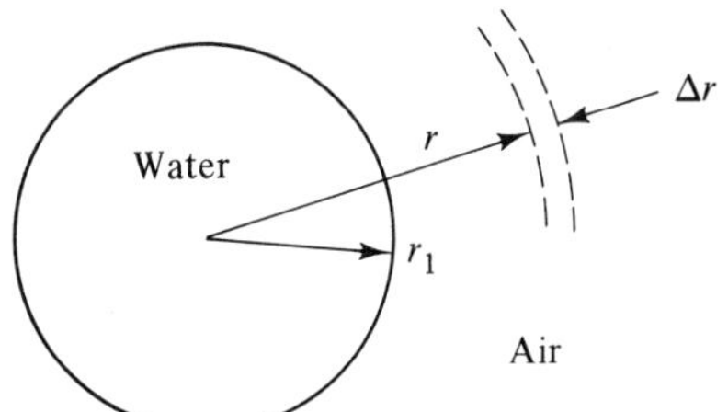

Figure 9.2 Evaporation of a raindrop

Chapter 10, where we discuss mass transfer and convection, we will find that the solutions developed here for these three problems actually represent a special instance of the more general situation involving both molecular diffusion and convective mass transfer.

9.6.1 Evaporation of a Spherical Droplet. The first problem we will consider is the evaporation of a droplet such as a raindrop or a droplet of oil in a furnace. The vapor formed at the surface of the droplet is assumed to move by molecular diffusion into the large body of stagnant gas that surrounds the droplet.

For a raindrop, the situation is as indicated in Figure 9.2. At any moment when the radius of the drop is r_1, the flux of water vapor at any distance r is given by equation (9.9).

$$N_w = -\frac{CD_{wa}}{1 - x_w}\frac{\partial x_w}{\partial r} \tag{9.9}$$

The flux N_w is not constant but, because of the spherical geometry, must decrease as the distance out from the surface of the droplet increases. What is constant, or at least nearly so, is the total molar flow rate of water vapor. Assuming steady-state conditions, this molar flow rate is the product of the flux and the area of the sphere through which the water vapor is passing. For the spherical shell of radius r and thickness Δr, this may be stated mathematically in the following way:

$$\text{Molar flow rate of water vapor} = |AN_w|_r = |AN_w|_{r+\Delta r}$$

where A = surface area of a sphere of radius r or $r + \Delta r$.

For steady conditions,

$$|4\pi r^2 N_w|_{r+\Delta r} - |4\pi r^2 N_w|_r = 0$$

or

$$\lim_{\Delta r \to 0} \frac{|r^2 N_w|_{r+\Delta r} - |r^2 N_w|_r}{\Delta r} \equiv \frac{d}{dr}(r^2 N_w) = 0 \tag{9.26}$$

Integrating equation (9.26), we obtain

$$r^2 N_w = \text{constant} \tag{9.27}$$

At the surface of the droplet,

$$r^2 N_w \big|_{r1} = r_1{}^2 N_{ws} \tag{9.28}$$

Combining equations (9.9), (9.27), and (9.28), we find

$$r_1{}^2 N_{ws} = -\frac{r^2 C D_{wa}}{1 - x_w} \frac{\partial x_w}{\partial r} \tag{9.29}$$

Equation (9.29) is readily integrated since the flux at the surface of the droplet is constant (for steady conditions).

$$r_1{}^2 N_{ws} \int_{r1}^{\infty} \frac{dr}{r^2} = -CD_{wa} \int_{x_{ws}}^{x_{w\infty}} \frac{dx_w}{1 - x_w}$$

or

$$r_1{}^2 N_{ws} \left(-\frac{1}{r} \right)_{r1}^{\infty} = CD_{wa} \ln(1 - x_w) \Bigg|_{x_{ws}}^{x_{w\infty}}$$

which simplifies to

$$N_{ws} = \frac{CD_{wa}}{r_1} \ln \frac{1 - x_{w\infty}}{1 - x_{ws}} \tag{9.30}$$

In a manner similar to that used in obtaining (9.21), equation (9.30) may be rewritten in terms of pressures and partial pressures.

$$N_{ws} = \frac{PD_{wa}}{RTr_1} \ln \frac{P_{a\infty}}{P_{as}} \tag{9.31}$$

We can use equation (9.31) to estimate the time required to evaporate the drop. To begin this calculation, a simple mass balance is used, which states that the rate at which water vapor leaves the droplet by diffusion must be the same for steady-state conditions as the rate of decrease of liquid water from the droplet. Thus, for a droplet of radius r,

$$\frac{\text{Moles of water diffusing}}{\text{unit time}} = 4\pi r^2 N_{ws}$$

where the flux N_{ws} is given by equation (9.31) for diffusion into an infinite stagnant medium.

$$\frac{\text{Moles of water leaving droplet}}{\text{unit time}} = -\frac{d}{dt}\left(\frac{4}{3}\pi r^3 C_{wl}\right) = -4\pi r^2 C_{wl} \frac{dr}{dt}$$

where C_{wl} = molar density of the liquid water.

Equating the rate at which water vapor diffuses away from the droplet to the rate of decrease of liquid in the droplet gives

$$4\pi r^2 N_{ws} = -4\pi r^2 C_{wl} \frac{dr}{dt}$$

which simplifies to

$$\frac{dr}{dt} = -\frac{N_{ws}}{C_{wl}} = -\frac{PD_{wa}}{C_{wl}RTr} \ln \frac{P_{a\infty}}{P_{as}} \tag{9.32}$$

Integrating,

$$\int_{r1}^{0} r\, dr = -\frac{PD_{wa}}{C_{wl}RT} \ln \frac{P_{a\infty}}{P_{as}} \int_{0}^{t} dt$$

where r_1 = initial radius of the droplet. Solving for the time of evaporation t, we obtain finally

$$t = \frac{r_1^2 C_{wl} RT}{2PD_{wa} \ln (P_{a\infty}/P_{as})} \tag{9.33}$$

It should be noted that the problem of an evaporating droplet is likely to be somewhat more complicated than discussed here. In the first place, instead of pure diffusion, the process is more likely to involve both diffusion and convection, with convection being more important for the larger drops. Another process that has been ignored in our discussion is the transfer of the heat that is required to generate the vapor at the surface of the droplet. At the beginning of the process, this heat comes from the droplet itself, which is quickly cooled to the adiabatic saturation temperature. The temperature of the droplet then remains constant at this value, requiring that the heat for evaporation be transferred from the air to the surface of the droplet by the processes of conduction and convection. Thus the problem of an evaporating drop is really one of simultaneous heat and mass transfer. In engineering operations such as humidification, dehumidification, and cooling of process water in a spray tower, the design of the equipment requires that this complex process of simultaneous heat and mass transfer be well understood.

9.6.2 Combustion of a Particle of Coal. If the particle of coal is assumed to be spherical (Figure 9.3), the problem of combustion is similar to evaporation of a droplet except that a chemical reaction occurs at the surface of the particle. This chemical reaction

$$C + O_2 \rightarrow CO_2$$

shows that for every mole of oxygen that diffuses to the surface to combine with the carbon, 1 mole of carbon dioxide is formed and must diffuse away from this surface. Thus this is a case of equimolar counterdiffusion of two gases, O_2 and CO_2. If we make the same simplifying assumptions as we did with the evaporating drop, the steady-state flux of O_2 to the particle may be expressed in the following way, applying equation (9.8):

$$N_{O_2} = -CD_{O_2\text{-gas}} \frac{\partial x_{O_2}}{\partial r} + x_{O_2} (N_{O_2} + N_{CO_2} + N_{N_2}) \tag{9.34}$$

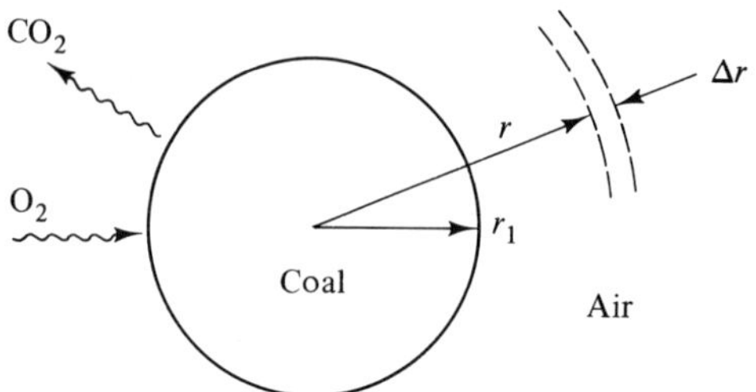

Figure 9.3 Combustion of a particle of coal

Since $N_{O_2} = -N_{CO_2}$ and $N_{N_2} = 0$, equation (9.34) reduces to

$$N_{O_2} = -CD_{O_2\text{-gas}} \frac{\partial x_{O_2}}{\partial r} \tag{9.35}$$

where $D_{O_2\text{-gas}}$ = diffusivity of O_2 in the gaseous mixture.

For steady conditions, the total molar flow rate of O_2 and of CO_2 is independent of the radius r. This is expressed in a material balance leading to a result that is similar to equation (9.26) or (9.27).

$$\frac{d}{dr}(r^2 N_{O_2}) = 0, \qquad \frac{d}{dr}(r^2 N_{CO_2}) = 0 \tag{9.36}$$

Thus, integrating equation (9.36) for the O_2 gives

$$r^2 N_{O_2} = \text{constant} = r_1^{\,2} N_{O_2s}$$

where the value of the constant has been evaluated at the surface of the particle. Using equation (9.35), the following expression is obtained:

$$r_1^{\,2} N_{O_2s} = -r^2 CD_{O_2\text{-gas}} \frac{\partial x_{O_2}}{\partial r} \tag{9.37}$$

This may be integrated as follows:

$$r_1^{\,2} N_{O_2s} \int_{r_1}^{\infty} \frac{dr}{r^2} = -CD_{O_2\text{-gas}} \int_{x_{O_2s}}^{x_{O_2\infty}} dx_{O_2}$$

which gives

$$N_{O_2s} = \frac{-CD_{O_2\text{-gas}}}{r_1}(x_{O_2\infty} - x_{O_2s}) \tag{9.38}$$

If the reaction at the surface of the particle takes place very quickly, the mole fraction of O_2 at the surface will approach zero. That is,

$$x_{O_2s} \approx 0$$

Also, at some distance from the particle, the mole fraction of oxygen will correspond to that of air.

$$x_{O_2\infty} = 0.21$$

Thus, with these conditions, equation (9.38) reduces to

$$N_{O_2s} = \frac{-0.21CD_{O_2\text{-gas}}}{r_1} \tag{9.39}$$

9.6.3 Oxidation of Sulfur Dioxide at a Catalytic Surface. The combustion of a particle of coal discussed in Section 9.6.2 involved two gases diffusing in opposite directions at equimolar rates. In the present problem of oxidation of SO_2, different gases are diffusing in opposite directions, but the rates of diffusion are not equimolar.

The chemical reaction taking place on the catalytic surface (Figure 9.4) is

$$SO_2 + \tfrac{1}{2}O_2 \rightarrow SO_3$$

Thus, for every mole of SO_3 formed and which must diffuse away from the surface, $1\frac{1}{2}$ moles of reactant gases must diffuse toward the catalytic surface.

Again, suppose the geometry of the catalytic particle is spherical. A material balance for each diffusing gas results in the following equations:

$$\frac{d}{dr}(r^2N_{O_2}) = 0, \qquad \frac{d}{dr}(r^2N_{SO_2}) = 0, \qquad \frac{d}{dr}(r^2N_{SO_3}) = 0 \tag{9.40}$$

which are obtained in precisely the same way as discussed earlier for the case of the evaporating droplet and combustion of a particle of coal. Integrating the expression for SO_2 gives

$$r^2N_{SO_2} = \text{constant} = r_1{}^2N_{SO_2s} \tag{9.41}$$

where the constant has been evaluated at the surface of the particle. The flux of SO_2 in accordance with equation (9.8) is

$$N_{SO_2} = -CD_{SO_2\text{-gas}}\frac{\partial x_{SO_2}}{\partial r} + x_{SO_2}(N_{SO_2} + N_{O_2} + N_{SO_3} + N_{N_2}) \tag{9.42}$$

Figure 9.4 Oxidation of SO_2 at the external surface of a catalyst pellet

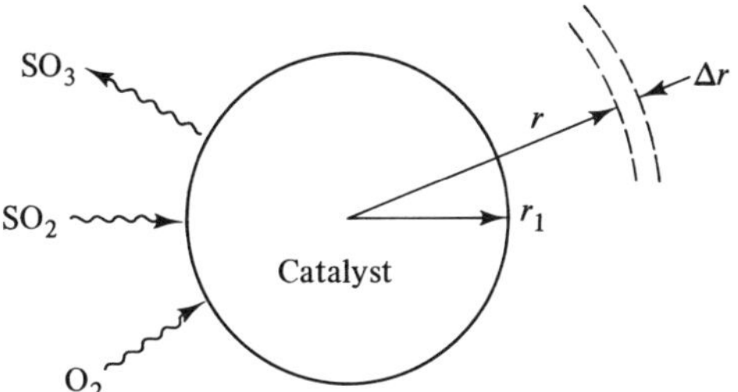

where the presence of N_2 comes about because air is used as the source of O_2. The N_2 is stagnant, $N_{N_2} = 0$; and for the stoichiometric reaction

$$N_{SO_2} = -N_{SO_3}, \qquad N_{O_2} = 0.5N_{SO_2}$$

Thus, equation (9.42) reduces to the following:

$$N_{SO_2} = -CD_{SO_2\text{-gas}}\frac{\partial x_{SO_2}}{\partial r} + 0.5x_{SO_2}N_{SO_2}$$

or

$$N_{SO_2} = \frac{-CD_{SO_2\text{-gas}}}{1 - 0.5x_{SO_2}}\frac{\partial x_{SO_2}}{\partial r} \tag{9.43}$$

Combining equations (9.41) and (9.43) and integrating gives

$$r_1^2N_{SO_2s}\int_{r_1}^{\infty}\frac{dr}{r^2} = -CD_{SO_2\text{-gas}}\int_0^{x_{SO_2\infty}}\frac{dx_{SO_2}}{1 - 0.5\,x_{SO_2}}$$

where it has been assumed that the reaction at the surface of the catalyst is occurring so quickly that the concentration of reactants at this surface has been reduced to zero.

The result of the integration is

$$N_{SO_2s} = \frac{2CD_{SO_2\text{-gas}}}{r_1}\ln(1 - 0.5\,x_{SO_2\infty}) \tag{9.44}$$

REFERENCES

1. Arnold, J. H., *Ind. Eng. Chem.*, *22* (10), 1091–95 (1930).
2. Perry, R. H., and C. H. Chilton, *Chemical Engineer's Handbook*, 5th ed. New York: McGraw-Hill Book Company, 1973.
3. Sherwood, T. K., and R. L. Pigford, *Absorption and Extraction*, 2nd ed. New York: McGraw-Hill Book Company, 1952.
4. Stefan, J., *Wiener Berichte*, *68*, 385–425 (1874). Referred to by E. R. G. Eckert and R. M. Drake in *Introduction to the Transfer of Heat and Mass*, pp. 236–40. New York: McGraw-Hill Book Company, 1950.

PROBLEMS

9.1. Calculate the value of the diffusivity of benzene in air at 101.3 kPa pressure and 0°C using the Gilliland correlation and compare with the value listed in Table A.6. The molecular weight of benzene is 78 and for air it is 28.9. *Answer:* 258 cm²/hr

9.2. Calculate the value of the diffusivity of carbon dioxide in nitrogen at 101.3 kPa pressure and 20°C using the Gilliland correlation and compare with the value listed in Table A.6. *Answer:* 482 cm²/hr

9.3. Using the Gilliland correlation, calculate the diffusivity of benzoyl chloride (C_6H_5COCl) in air at 40°C and 101.3 kPa pressure. *Answer:* 255 cm²/hr

9.4. A glass cylinder, closed at the bottom and open at the top, contains a layer of liquid water at the bottom. Dry air at 20°C and 101.3 kPa pressure passes slowly across the open end of the cylinder. Calculate the flux of water vapor from the cylinder for steady-state conditions if the diffusion distance is 25 cm. Assume the water temperature is 20°C, at which temperature its vapor pressure is 17.5 mm Hg. *Answer:* 3.41×10^{-4} kg mol/hr-m²

9.5. An open circular tank 4.6 m in diameter contains benzene at 20°C, which is exposed to the atmosphere. If benzene is worth 40 cents per liter, what is the value of the benzene lost from the tank in dollars per day? Assume a 5-mm-thick layer of stagnant air at the surface of the benzene. The vapor pressure and density of benzene at 20°C are 95 mm Hg and 0.88 g/ml, respectively. *Answer:* $490/day

9.6. One gram of iodine is placed at the bottom of a vertical tube that is 25 cm long and 4-mm inside diameter. How long will it take for the solid iodine to disappear if the tube and its contents are at room temperature (20°C) and 1 atm pressure. The vapor pressure of iodine at 20°C is 0.202 mm Hg. *Answer:* 28.5 years.

9.7. A moth ball is a solid sphere of naphthalene about 13 mm in diameter. How long would it take for such a sphere to disappear if placed in stagnant air at room temperature (20°C)? The density and vapor pressure of naphthalene at 20°C are 1145 kg/m³ and 0.040 mm Hg, respectively. *Answer:* 0.43 year

9.8. A vertical pipe (25-mm ID), 3 m long, and closed at the bottom contains a 15-cm-deep layer of water at the bottom. How long would it take for this water to evaporate:

(a) Assuming that the air in contact with the open end of the pipe is dry?

(b) Assuming that the air has a relative humidity of 70%?

The temperature and pressure of the ambient air are 21°C and 1 atm, respectively. The vapor pressure of water at 21°C is 19 mm Hg. *Answer:* (a) 29 years; (b) 97 years

9.9. A thin porous metal plate, permeable to CO only and with one surface coated with a combustion catalyst, is placed near the end of a long cylindrical duct as illustrated in the sketch. One side of the plate is in contact with a supply of carbon monoxide, and the other side, which is coated with the combustion catalyst, is in contact with the surrounding air. Both gases are at 38°C and 1 atm pressure. Estimate the maximum steady-state flow of carbon monoxide to the plate if complete combustion of the gas is maintained at the catalyst surface. Assume the transfer of oxygen to the catalytic surface is by molecular diffusion only. *Answer:* 4.0×10^{-4} kg mol/m²-hr

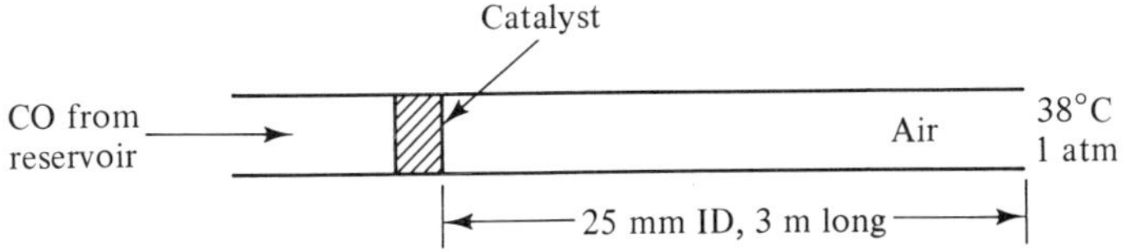

9.10. A droplet of water 5 mm in diameter is suspended in still air that has a temperature of 21°C, a pressure of 1 atm, and a relative humidity of 50%. How long will it take for the droplet to evaporate completely? The vapor pressure of water at 21°C is 18.65 mm Hg. *Answer:* 3.8 hr

9.11. A solid sphere of carbon, 5 mm in diameter, is suspended in still air at 1 atm pressure and 540°C. Estimate the time it would take for the sphere to oxidize completely to carbon dioxide. List all the assumptions upon which this calculation is based. The density of carbon is 2.25 g/ml. *Answer:* 0.55 hr

9.12. A spherical pellet of vanadium pentoxide catalyst is suspended in a chamber held at 450°C. The chamber is filled with a mixture of sulfur dioxide (8% by volume) and air at 1 atm pressure. Assuming the chamber to be large compared to the pellet and that no convection flow occurs in the gas in the chamber, estimate the rate at which sulfur trioxide is formed at the surface of the pellet. The pellet radius is 4 mm. *Answer:* 0.22 g/hr

Convective Mass Transfer 10

Our discussion of mass transfer in Chapter 9 was limited to molecular diffusion, which we learned is a process resulting from a concentration gradient. In systems involving liquids or gases, however, we know it is very difficult to eliminate convection from the overall mass-transfer process. For instance, in the case of an evaporating droplet, it would be almost impossible to eliminate completely bulk motion in the air surrounding the droplet. As a consequence, the calculated time to evaporate the droplet, based on the assumption that the water vapor diffuses into the surrounding air that is stagnant, would be seriously in error. In the more realistic situation, where convection is coupled to molecular diffusion, the effect is to decrease considerably the overall resistance to mass transfer and, therefore, to increase the rate and decrease the time required to evaporate the droplet to dryness.

Our earlier discussion was also limited to diffusion within a single phase. Most applications of mass transfer, however, involve transfer between phases. A droplet of water evaporating into air, for instance, involves two phases, with the mass transfer occurring from the interface into the air. There is no mass transfer within the droplet itself since it is a pure liquid. If, however, a component of air, say oxygen, was being absorbed by the droplet, then mass transfer would take place in both the liquid and gaseous phases. Our discussion here will consider the simpler situation first, that of mass transfer within a single phase either to or from the phase boundary. We will then follow this by discussion of the more complex case of two fluids in contact, with mass transfer taking place in both phases.

10.1 Mass Transfer within a Single Fluid Phase

If a sheet or droplet of pure liquid evaporates into the surrounding air, mass transfer occurs only from the phase boundary to the body of air. Similarly, if a pure solid dissolves in a liquid or sublimes into a gas, mass transfer occurs only in the liquid or the gas. A metal electrode discharging cations from a solution also involves mass transfer in the liquid phase only, but in this case the direction of mass transfer is from the bulk liquid to the phase boundary at the surface of the electrode.

The engineering design of systems involving such processes requires that we express the rate of mass transfer mathematically. This in turn requires that we have a mechanistic picture or model of the transfer process. The fact that the resistance to mass transfer in a turbulent fluid is largely confined to a very thin layer adjacent to the phase boundary led investigators in the early part of this century to the concept of a stagnant film at the phase boundary through which the transfer takes place. The notion of a stagnant film is a simplification of conditions at the interface but nevertheless has been widely used in defining the rate of mass transfer.

Figure 10.1 illustrates conditions near the phase boundary when a fluid F moves over a plane surface of pure solid. A transfer of s takes place from the phase boundary to the bulk fluid under a concentration gradient established adjacent to the boundary (s must be soluble if F is a liquid or sublime if F is a gas). The steady-state rate of mass transfer may be expressed as in equation (1.4).

$$\text{Rate of mass transfer} = \frac{\text{potential}}{\text{resistance}} \tag{1.4}$$

where the potential is defined as the difference in concentration of s across the mass-transfer boundary layer:

$$\text{Potential} = \Delta c_s = c_{si} - c_{sF} \text{ mol of s/m}^3 \tag{10.1}$$

Figure 10.1 Mass transfer at the interface between a pure solid and a fluid

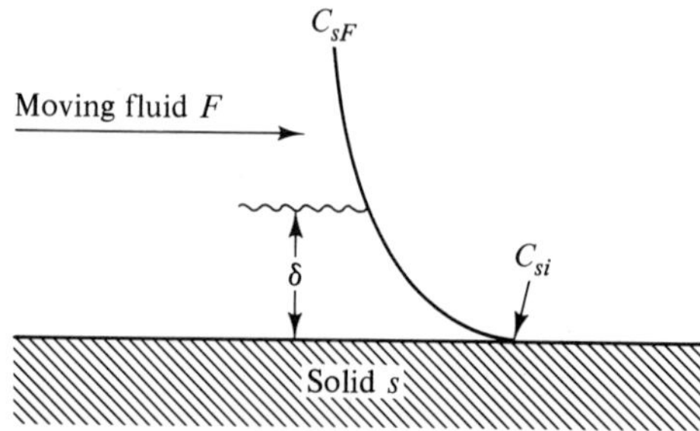

where c_{si} = concentration of s at the solid–fluid interface, mol of s/m^3

c_{sF} = concentration of s in the bulk fluid (c_{sL} if F is a liquid, c_{sG} if F is a gas), mol of s/m^3

The resistance in equation (1.4) is expressed in the same way as was done in the analogous heat transfer case, that is, in terms of a coefficient k and the area A through which the mass transfer occurs:

$$\text{Resistance} = \frac{1}{k_c A} \tag{10.2}$$

where k_c = mass transfer coefficient, the subscript c referring to the potential expressed in terms of concentrations

A = area that is normal to the direction of mass transfer, m^2.

The rate of mass transfer is thus written as

$$m_s = k_c A (c_{si} - c_{sF}) \tag{10.3}$$

where m_s = molar mass transfer rate, mol of s/s.
An alternative way of writing equation (10.3) is

$$N_s = k_c (c_{si} - c_{sF}) \tag{10.4}$$

where $N_s = m_s/A$, the molar flux of s, mol of s/m^2-s.

The units of k_c as derived from equation (10.3) or (10.4) are seen to be m/s.

Other ways of expressing the potential in equation (1.4) are often used, such as mole fraction or, when fluid F is a gas, partial pressure. For instance, when F is a liquid, the mole fraction is represented by x, and the rate equation is written in the following way:

$$N_s = k_x (x_{si} - x_{sL}) \tag{10.5}$$

where x_{si}, x_{sL} = mole fraction of s in the liquid at the solid–liquid interface and in the bulk liquid, respectively. If F is a gas, either the mole fraction y or the partial pressure p may be used in place of the concentration c in the rate equation, giving

$$N_s = k_y (y_{si} - y_{sG}) \tag{10.6}$$

$$N_s = k_p (p_{si} - p_{sG}) \tag{10.7}$$

where the subscripts i and G refer to the solid–gas interface and the bulk gas, respectively.

When applying equations such as (10.5), there is usually no difficulty in defining the concentration in the bulk fluid phase since this is readily measured. The concentration at the solid–fluid interface, however, is not readily defined except by making an assumption about conditions at this interface.

The assumption that is usually made is that the two phases in contact at the interface are in equilibrium with each other. This means that data such as solubilities and vapor pressures can be used to define the concentration at the interface, as illustrated in the following two examples.

Example 10.1.1

A slab of sodium chloride (NaCl), measuring 10 cm by 10 cm with edges sealed, is suspended in a well-stirred tank of water. It is found to be losing weight at the rate of 40.8 g/min. Estimate the mass transfer coefficient k_c, given that the solubility of sodium chloride in water is 36.0 g/100 g of water at the temperature in the tank (20°C).

The density of a saturated solution of NaCl at 20°C is 1.1972 g/cm³. Hence

$$\Delta C_B = \frac{36.0}{136.0} \times 1.1972 - 0 = 0.317 \text{ g/cm}^3$$

and

$$N_B = 40.8 \frac{\text{g}}{\text{min}} \times \frac{\text{min}}{60 \text{ s}} \times \frac{1}{2 \times 10 \times 10 \text{ cm}^2} = 3.4 \times 10^{-3} \text{ g/s-cm}^2$$

Hence

$$k_c = \frac{N_B}{\Delta C_B} = \frac{3.4 \times 10^{-3}}{0.317} = 1.1 \times 10^{-2} \text{ cm/s}$$

Example 10.1.2

A slab of naphthalene, measuring 4 cm by 4 cm with all four edges and the back surface sealed, is suspended in the air of a large room. It is found to be losing weight at the rate of 2.0 g/day. Estimate the mass transfer coefficient k_p between naphthalene and air. The sublimation pressure of naphthalene at the temperature of the room (25°C) is 0.040 mm Hg.

From the given data

$$\Delta p_B = p_{Bi} - p_{BG} = 0.040 - 0 = 0.040 \text{ mm Hg}$$

$$= 0.040 \times 133.3 \text{ Pa}$$

and

$$N_B = \frac{2.0}{24} \frac{\text{g}}{\text{hr}} \times \frac{(100)^2}{4 \times 4\text{m}^2} \times \frac{1}{128.16} = 0.41 \text{ g mol/m}^2\text{-hr}$$

Hence

$$k_p = \frac{0.41 \text{ mol}}{\text{m}^2\text{-hr}} \times \frac{1}{0.040 \times 133.3 \text{ Pa}} \times \frac{\text{hr}}{3600 \text{ s}}$$

$$= 2.1 \times 10^{-5} \text{ g mol/m}^2\text{-s-Pa}$$

10.1.1 Relationship between k and the Diffusivity D. If the film at the interface shown in Figure 10.1 is indeed stagnant, as assumed in the stagnant-film model, solute s must move from the phase boundary to the bulk fluid by molecular diffusion. The flux expressed by equation (10.4) can then be equated to the molecular diffusion flux given by (9.20) to obtain a relationship between k_c and the diffusivity D_{sF}.

$$N_s = k_c(c_{si} - c_{sF}) = \frac{D_{sF}}{\delta(x_F)_{1m}}(c_{si} - c_{sF}) \tag{10.8}$$

whence

$$k_c = \frac{D_{sF}}{\delta(x_F)_{1m}} \tag{10.9}$$

The coefficient k_c(or k_x, k_y, k_p) is thus seen to be directly proportional to D_{sF}. There is abundant evidence in the literature, however, to indicate that such a direct proportionality is probably not valid. This has led to many attempts to devise a more satisfactory model of the mass-transfer process. The usefulness of relationships such as equation (10.9) is limited in another respect as well. The distance parameter δ, the effective film thickness, cannot be experimentally measured or predicted from any theoretical relationship, at least not at the present time.

10.1.2 Other Models of the Mass-Transfer Process. The search for a more satisfactory expression of the mass-transfer process has gone on for many years, with the result that several new models of the process have been proposed. These will not be discussed in detail here since more advanced treatments of the subject have considered them recently (5, 10, 13). The model proposed by Higbie (3) will be discussed briefly, however, since it appears to be one of the earliest attempts to find an alternative to the stagnant-film concept and may as well have served as the spark to the imagination of other investigators, such as Danckwerts (1).

Higbie pictured a small fluid element moving from the bulk phase to the interface where it remains for a time period t, after which it returns to the bulk phase. During the time interval t, transient diffusion is assumed to take place between the interface and the element, resulting in the following relationship between the mass-transfer coefficient and the diffusivity:

$$k_c = 2\left(\frac{D_{sF}}{\pi t}\right)^{1/2} \tag{10.10}$$

The square-root relationship between k_c and D_{sF} may be more reasonable than the direct proportionality suggested by the stagnant-film model. However, the time t is not known for situations of practical interest. Higbie's model is therefore deficient in a way similar to the stagnant-film idea, where the thickness δ is an unknown parameter.

In spite of its deficiencies, the stagnant-film model is still used as the basis for correlations of mass-transfer data. This will become apparent in the discussion to follow.

10.2 Mass-Transfer Coefficient at Surfaces of Simple Geometry

In our earlier discussion of convective heat transfer, it seemed reasonable to suppose that the coefficient h must depend on certain properties of the fluid, as well as on the dynamics and geometry of the flow (see Section 6.6). We also found that dimensional analysis was a useful starting point in our search for a mathematical relationship between the variables involved. The similarity between convective mass transfer and convective heat transfer suggests that the same type of approach might be used in developing correlations of mass-transfer data.

10.2.1 Mass Transfer at a Plane Surface. Consider the situation illustrated in Figure 10.1, where a fluid F moves by forced convection over a plane surface s. Suppose the coefficient k_c depends on the following variables: fluid density ρ, fluid viscosity μ, diffusivity D_{sF}, average fluid velocity V, and the surface length in the direction of flow L. We may therefore write

$$k_c = f(\rho, \mu, D_{sF}, V, L) \tag{10.11}$$

which is the starting point for dimensional analysis. We proceed with the analysis in the same way as we did in Section 3.4 when we were discussing pipe friction. Equation (10.11) is rewritten as an infinite power series, only one term of which we need to consider:

$$k_c = C\rho^a\mu^b D_{sF}{}^c V^d L^e \tag{10.12}$$

where C is a dimensionless constant. Substituting the dimensions for the various quantities using the SI system, the dimensional form of equation (10.12) becomes

$$Lt^{-1} \underset{=}{D} (ML^{-3})^a (ML^{-1}t^{-1})^b (L^2t^{-1})^c (Lt^{-1})^d (L)^e \tag{10.13}$$

where the symbol $\underset{=}{D}$ means *dimensionally equal*. We now write an equation for each dimension, which is a statement that the dimensions on both sides of equation (10.13) must be the same. Thus

$$M \qquad 0 = a + b \tag{10.14}$$

$$L \qquad 1 = -3a - b + 2c + d + e \tag{10.15}$$

$$t \qquad -1 = -b - c - d \tag{10.16}$$

These three equations can be solved for any three of the unknown powers in terms of the other two. Choosing to solve for a, b, and e in terms of c and d, the result is

$$a = -b = -1 + c + d$$

From equation (10.14) and (10.16), and from (10.15) with substitutions,

$$\begin{aligned} e &= 1 + 3a + b - 2c - d \\ &= 1 + 2a - 2c - d \\ &= 1 + 2(-1 + c + d) - 2c - d \\ &= -1 + d \end{aligned}$$

Substituting for a, b, and e in equation (10.12), we obtain

$$k_c = C\rho^{-1+c+d}\mu^{1-c-d}D_{sF}^{c}V^{d}L^{-1+d}$$

or

$$k_c = C\left(\frac{\mu}{\rho L}\right)\left(\frac{\rho D_{sF}}{\mu}\right)^{c}\left(\frac{LV\rho}{\mu}\right)^{d}$$

or

$$\frac{k_c\rho L}{\mu} = f_1\left(\frac{LV\rho}{\mu}, \frac{\mu}{\rho D_{sF}}\right) \tag{10.17}$$

where the two dimensionless groups within parentheses are recognized as the Reynolds number, Re_L, and Schmidt number, Sc, respectively. The group on the left side of equation (10.17) is also dimensionless but is not usually used when correlating data. The usual form is obtained by multiplying (10.17) by the Schmidt number, giving

$$\frac{k_cL}{D_{sF}} = f\left(\frac{LV\rho}{\mu}, \frac{\mu}{\rho D_{sF}}\right) \tag{10.18}$$

The dimensionless group on the left of equation (10.18) is sometimes referred to as the Nusselt number for mass transfer, but by many authors as the Sherwood number, Sh_L. Thus (10.18) may be written as

$$\mathrm{Sh}_L = f(\mathrm{Re}_L, \mathrm{Sc}) \tag{10.19}$$

indicating the three dimensionless groups required to correlate mass-transfer data for flow over a plane surface. The function f is, of course, not revealed by dimensional analysis but will be defined empirically if equation (10.19) is shown to be successful in correlating the data.

Most of the data on mass transfer from flat plates have been obtained for the evaporation of various liquids or the sublimation of solids such as naphthalene into air. An early correlation of these data by Sherwood and Pigford (9) was modified by Sissom and Pitts (13). Figure 10.2 illustrates this correlation. It can be seen that a single straight line is a fair representation of all the data points on this figure. The equation of this line, given

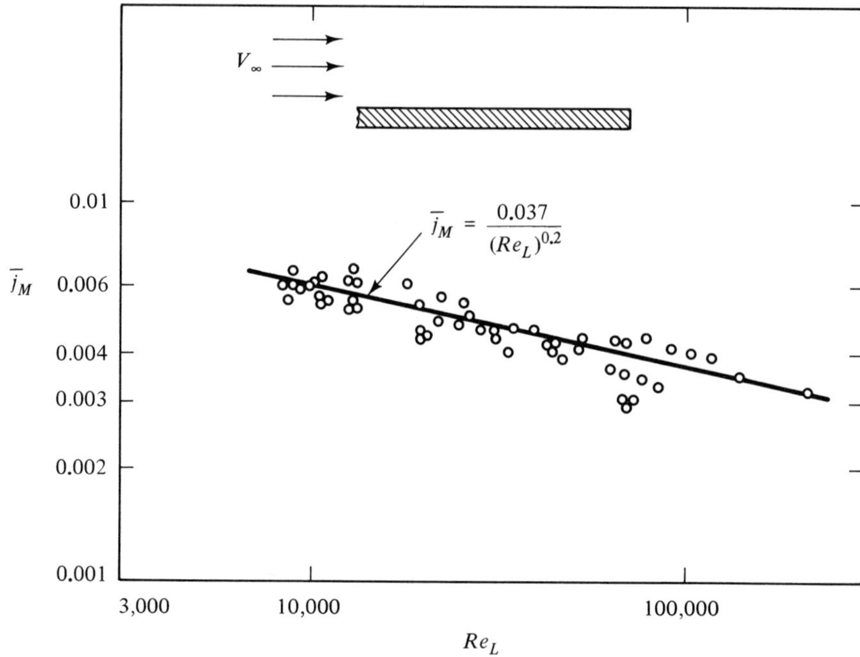

Figure 10.2 Mass transfer from a flat plate for Sc > 0.6 (13)

by Sissom and Pitts (13), is

$$\bar{j}_M = \frac{0.037}{\mathrm{Re}_L^{0.2}} \tag{10.20}$$

where $\bar{j}_M = k_c \mathrm{Sc}^{2/3}/V$. Equation (10.20) is readily rearranged to the following form:

$$\mathrm{Sh}_L = 0.037(\mathrm{Re}_L)^{0.9}(\mathrm{Sc})^{1/3} \tag{10.21}$$

We can therefore say that, within the limitations of the data and its correlation as illustrated in Figure 10.2, the function f of equation (10.19) is defined by equation (10.21).

Example 10.2.1

A backyard pool, 10 m long by 8 m wide, has an average temperature of 22°C. Estimate the rate of evaporation from the pool on a day when the wind speed is 1 m/s, the air temperature is 22°C, and the relative humidity is 70%. The vapor pressure of water at 22°C = 19.8 mm Hg.

The Reynolds number is

$$\mathrm{Re}_L = \frac{LV\rho}{\mu} = \frac{10 \times 1}{1.6 \times 10^{-5}} = 6.3 \times 10^5$$

where it is assumed that the wind direction is parallel to the length of the pool and the pool surface acts as a flat plate. From equation (10.20),

$$\bar{j}_M = \frac{0.037}{(6.3 \times 10^5)^{0.2}} = 0.0026$$

The Schmidt number is

$$\text{Sc} = \frac{\mu}{\rho D_{sF}} = \frac{1.6 \times 10^{-5}}{2.5 \times 10^{-5}} = 0.64$$

where the diffusion coefficient for water vapor in air at 22°C is estimated to be 2.5×10^{-5} m²/s. Thus

$$k_c = \frac{\bar{j}_M V}{\text{Sc}^{2/3}} = \frac{0.0026 \times 1}{0.64^{2/3}} = 0.0035 \text{ m/s}$$

The rate of evaporation is given by equation (10.3):

$$m_s = k_c A(c_{si} - c_{sG}) \tag{10.3}$$

Assuming the ideal gas law is valid for water–air mixtures gives

$$c_{si} = \frac{n_i}{V} = \frac{P_i}{RT} = \frac{19.8}{760} \times \frac{1.0133 \times 10^5}{8314 \times 295} \text{ kg mol/m}^3$$

and

$$c_{sG} = \frac{0.70 \times 19.8}{760} \times \frac{1.0133 \times 10^5}{8314 \times 295} \text{ kg mol/m}^3$$

Hence

$$m_s = 0.0035 \times 10 \times 8(1 - 0.70)\frac{19.8}{760} \times \frac{1.0133 \times 10^5}{8314 \times 295}$$

$$= 9.0 \times 10^{-5} \text{ kg mol/s}$$

$$= 9.0 \times 10^{-5} \times 18 \times 3600 = 5.8 \text{ kg/hr}$$

$$= 5.8 \times \frac{1000}{453.6} \times \frac{1}{10} \times 24 = 31 \text{ Imperial gal/day}$$

10.2.2 Mass Transfer at the Surface of a Sphere. Our earlier discussion of this case (Section 9.6) assumed the surrounding fluid to be stagnant so that mass transfer occurred strictly by the process of molecular diffusion. Referring again to Figure 9.3 and using subscript s for the sphere and F (or L, G) for the surrounding fluid, we will first relate the mass-transfer coefficient k_c to this earlier discussion.

For steady conditions, the molar flow rate of the diffusing substance s is the same in the fluid at any distance r [equation (9.27)]. Applying this to the surface of the sphere of radius r_1 and to the surface at distance r, we have

$$4\pi r_1^2 N_{s1} = 4\pi r^2 N_s \tag{9.27}$$

where N_{s1}, N_s = flux of s at radii r_1 and r, respectively. The flux N_s is given by Fick's equation (9.1) if the convective flux due to diffusion is neglected for the sake of simplicity:

$$N_s = -D_{sF}\frac{dc_s}{dr} \tag{10.22}$$

Substituting for N_s in equation (9.27) from (10.22) gives

$$r_1{}^2 N_{s1} = -D_{sF}r^2\frac{dc_s}{dr} \tag{10.23}$$

Integration between the limits $c_s = c_{si}$ at $r = r_1$ and $c_s = c_{s\infty}$ at $r = \infty$ gives

$$r_1{}^2 N_{s1}\int_{r_1}^{\infty}\frac{dr}{r^2} = -D_{sF}\int_{c_{si}}^{c_{s\infty}} dc_s$$

or

$$r_1 N_{s1} = D_{sF}(c_{si} - c_{s\infty}) \tag{10.24}$$

Applying equation (10.4) at the surface of the sphere gives

$$N_{s1} = k_c(c_{si} - c_{s\infty}) \tag{10.4}$$

Substituting for the concentration difference in equation (10.24) from (10.4), we obtain

$$r_1 N_{s1} = \frac{D_{sF}N_{s1}}{k_c}$$

or

$$\frac{k_c D_1}{D_{sF}} = 2 \tag{10.25}$$

where D_1 = sphere diameter = $2r_1$.

Since equation (10.25) is based on molecular diffusion into a fluid where there is no convection, the value of k_c derived from this equation represents a limiting value. Any bulk movement in the fluid will result in an increase in the rate of mass transfer, which will be reflected in a higher value for the coefficient k_c than given by equation (10.25).

The effect of convection on mass transfer from single spheres has been studied by many investigators both theoretically and experimentally. A recent review of this work is that by Sherwood et al. (11). The curve of Brian and Hales (12), reproduced here as Figure 10.3, is an excellent illustration of the effect of convection on the coefficient k_c. At low flows, represented in the figure by the Peclet number, Pe, which is the product of Re and Sc, the Sherwood number has the limiting value of 2 as given by equation (10.25). For higher values of Pe, the value of Sh increases, eventually becoming linear

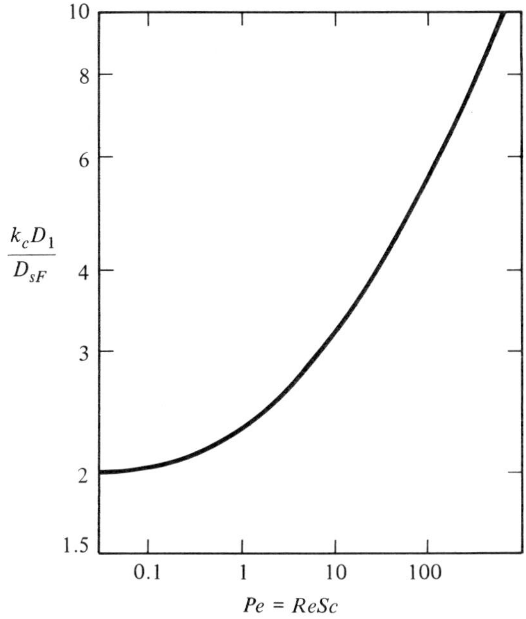

Figure 10.3 Mass transfer from a single sphere (plot by P. L. T. Brian and H. B. Hales, *AIChEJ*, *15*, 419, 1969. Reproduced by permission of the American Institute of Chemical Engineers)

on the log–log coordinates of Figure 10.3. Analytical expressions for the curve in this figure are given by Sherwood et al. (11):

$$\frac{k_c D_1}{D_{sF}} = (4.0 + 1.21\ \mathrm{Pe}^{2/3})^{1/2} \tag{10.26}$$

for Pe < 10,000, and

$$\frac{k_c D_1}{D_{sF}} = 1.0\mathrm{Pe}^{1/3} \tag{10.27}$$

for Pe > 1000.

The following problem is an interesting application of mass transfer to a sphere.

Example 10.2.2

A hailstone falling freely through the atmosphere grows by transfer of water vapor from the air. Estimate the coefficient k_c for a 3-cm hailstone that has a Re number of about 4×10^4.

Assume an air temperature of -10°C. The kinematic viscosity of air is

$$\nu = 1.35 \times 10^{-5}\ \mathrm{m^2/s}$$

The diffusivity of water vapor in air $= D_{sF} = 0.0792$ m²/hr. Hence the Peclet number is

$$\text{Pe} = \text{Re} \times \text{Sc} = 4 \times 10^4 \times \frac{1.35 \times 10^{-5}}{0.0792} \times 3600 = 2.5 \times 10^4$$

From equation (10.27),

$$k_c = 1.0\text{Pe}^{1/3}\left(\frac{D_{sF}}{D_1}\right) = 1.0(2.5 \times 10^4)^{1/3}\left(\frac{0.0792 \times 100}{3}\right)$$

$$= 77 \text{ m/hr}$$

According to Schuepp (7), who was interested in the growth of hailstones and in the variation of k_c around a sphere, this value of k_c is too low. His correlation is

$$\text{Sh} \times \text{Sc}^{-0.33} = 0.51\,\text{Re}^{0.5} + 0.02235\,\text{Re}^{0.78}$$

which predicts the value of k_c to be

$$k_c = \frac{0.0792 \times 100}{3}\left(\frac{1.35 \times 10^{-5} \times 3600}{0.0792}\right)^{0.33} [0.51(4 \times 10^4)^{0.5} + 0.02235(4 \times 10^4)^{0.78}] = 424 \text{ m/hr}$$

Such a large difference in the calculated value of k_c is certainly unacceptable, but is not readily explained at the present time.

10.3 Mass Transfer under Conditions of Natural Convection

In our discussion of mass transfer from a plate and a sphere, it was assumed that the flow past these objects was by forced convection. However, this need not always be the case. For instance, consider a solid sphere immersed in a large body of liquid that is stagnant. If the sphere is soluble in the liquid, dissolution will occur and the solution in contact with the sphere will have a density and therefore a specific weight that is different from that in the main body of liquid. This difference in specific weight gives rise to an unbalanced force, which establishes a flow around the sphere called *natural convection*. The mechanism of this flow is, of course, exactly the same as found in the analogous case of heat transfer (see Section 6.7), where the specific weight difference and resulting convection current are the result of a difference in temperature in the fluid.

Our earlier discussion of natural convective heat transfer (Section 6.7) showed that the Nusselt number for heat transfer was a function of the Prandtl and the Grashof number. We would therefore expect that for natural convective mass transfer the Nusselt number for mass transfer (or the Sherwood number) should be a function of the Schmidt number, which replaces the

Prandtl number, and the Grashof number. Thus

$$\mathrm{Sh} = f_1(\mathrm{Sc}, \mathrm{Gr}) \tag{10.28}$$

where Gr = dimensionless Grashof number = $(g\Delta\rho D^3)/\rho\nu^2$

g = acceleration of gravity

$\Delta\rho$ = difference in density between the saturated solution and the bulk fluid

D = sphere diameter

ρ, ν = density and kinematic viscosity of the fluid, respectively, evaluated at the mean film concentration

Several investigators have studied mass transfer from solids of simple shape to determine the function f_1 in equation (10.28), as well as other details of the flow. We will consider two such cases, that of mass transfer from a single sphere and from a vertical plate.

10.3.1 Single Sphere. Schütz (8) studied natural convective mass transfer from a single sphere by means of an electrochemical technique. In this method, a nickel-plated sphere was used as the cathode for the deposition of copper ions from an aqueous solution of copper sulfate and sulfuric acid. By electrically isolating a small portion of the surface of the sphere, he was also able to determine the local coefficient of mass transfer and how it varied with position around the sphere. His measurements of the average coefficient k_c were well correlated by the following equation:

$$\mathrm{Sh} = 2 + 0.59\,(\mathrm{Gr\,Sc})^{1/4} \tag{10.29}$$

over the range $2 \times 10^8 < \mathrm{Gr\,Sc} < 1.5 \times 10^{10}$. The number 2 in this equation is the limiting value of the Sherwood number for diffusion into an infinite medium as given by equation (10.25).

Another method of studying the same phenomena is illustrated by the work of Schenkels and Schenk (6). These investigators prepared solid spheres of different organic acids and immersed them in different solvents. From the rate of change in the diameter of the spheres measured photographically, they were able to calculate rates of mass transfer. Their correlation of the data is similar to equation (10.29) given by Schütz. Of special interest is the flow pattern about the sphere made visible in photographs by illuminated dust particles and by dye threads. One of these photographs shows the formation of waves associated with instability in the boundary layer. Figure 10.4 is a sketch which illustrates this phenomenon. Such instability is believed to result in an increase in the rate of mass transfer from the sphere.

10.3.2 Vertical Plate. Several investigators have studied mass transfer from vertical plates under conditions of natural convection using either the electrochemical method or dissolution of castings of organic acids. The work

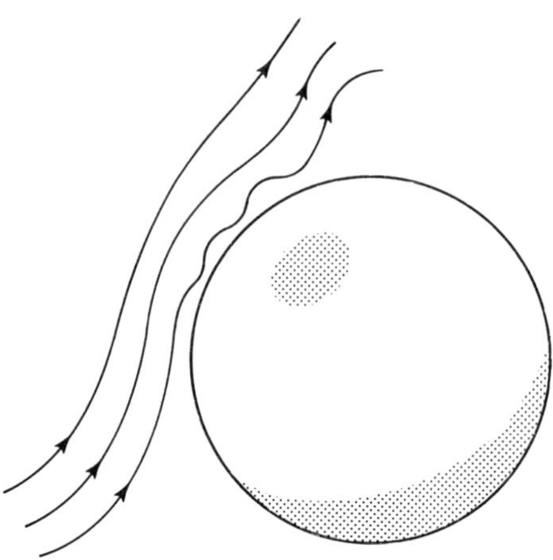

Figure 10.4 Sketch illustrating the origin of waves in the boundary layer flow about a sphere (6)

of Wilke et al. (16) is interesting in this regard since these authors used both methods and were able to obtain a single correlation for all their data. Their correlation is given by the equation

$$\mathrm{Sh} = 0.66(\mathrm{Sc} \times \mathrm{Gr})^{1/4} \tag{10.30}$$

which is clearly of the form of equation (10.28). The Grashof number in (10.30) is defined using the plate height L; that is,

$$\mathrm{Gr} = \frac{g\Delta\rho L^3}{\rho\nu^2}$$

The size of the plates used in developing the preceding correlation was quite small, ranging from 0.634 to 7.6 cm in the case of the electrochemical experiments and from 2.54 to 14 cm for the dissolution runs. Much larger plates were used by Husband (4), who was interested in the solution mining of potash. In his experimental work, plates of invert sugar were used where the plate height was varied from 7.6 cm to 6.1 m. With such high plates, Husband was able to demonstrate that the flow changed from laminar to turbulent, with the transition occurring at about $\mathrm{Gr} \times \mathrm{Sc} = 6 \times 10^{13}$. His correlation for the laminar region was

$$\mathrm{Sh} = 1.21(\mathrm{Gr} \times \mathrm{Sc})^{0.26} \tag{10.31}$$

and for the turbulent region

$$\mathrm{Sh} = 0.31(\mathrm{Gr} \times \mathrm{Sc})^{0.31} \tag{10.32}$$

Both correlations are clearly of the form of equation (10.28), but differ from the equation of Wilke et al. (16) in both the constant and the power on the group $\mathrm{Gr} \times \mathrm{Sc}$. There appear to be two principal reasons for this, the first being the low concentrations used in the Wilke study and, second, the roughness developed on the much more soluble surfaces used by Husband.

Our brief discussion of natural convective mass transfer about a sphere and a vertical plate has given some insight into this mode of material transfer, as well as the methods used to correlate data on rates of mass transfer. The subject would seem, nevertheless, to be of minor interest to engineers as witnessed by a recent textbook on mass transfer (10), which does not even mention the topic. There is one engineering application, however, where natural convective mass transfer is of major importance in the rate of the overall process. This is the solution mining of soluble ores such as salt (NaCl) or potash (KCl). As mentioned previously, the work of Husband (4) on tall vertical surfaces was carried out to increase our understanding of this process. A description of this interesting and unusual method of mining is pertinent to the present discussion and will be outlined briefly in the following section.

10.3.3 Solution Mining of Potash. Solution mining is the technique by which soluble ores are extracted at considerable depth from the earth's surface. In fact, the deeper lies the formation to be mined, the more economical does solution mining become compared to conventional methods of extracting the ore. In Western Canada, the formations mined by this method for potash lie over 900 m (3000 ft) below the surface of the earth.

The method used in this type of mining is illustrated in Figure 10.5. A hole is drilled to the bottom of the formation and fitted with two concentric pipes. The inner pipe is extended to the bottom of the formation, while the end of the outer pipe terminates some distance above this but below the insoluble interface. When water is passed into the formation, it dissolves the ore to form a solution that is removed at the bottom by the inner pipe. Initially, the water is injected at a rate so as to wash out a vertical cavity surrounding the pipes and extending over the total depth of the formation. The rate of water injection is then reduced so that the solution collecting at the bottom of the cavity and removed by the inner pipe is close to the saturation value. As the cavity increases in size exposing a larger area to the injected water, the rate of dissolution of the ore and hence the rate of production from the well increases.

The shape that the growing cavity assumes is a direct reflection of the variation in concentration of the solution over the depth of the cavity. Since the solution at the bottom of the cavity has the highest density and hence the largest concentration of potash, the rate of dissolution of the ore is smallest in this region. The injected water, having the smallest density, rises very quickly to the top of the cavity. When it makes contact with the ore, it contains very little dissolved potash and therefore dissolves the ore at the highest rate. The result is that the ore is dissolved most rapidly at the top giving a bowllike or morning-glory shape to the cavity.

One engineering problem associated with this process is to be able to predict the rate at which the cavity increases in size. This requires an under-

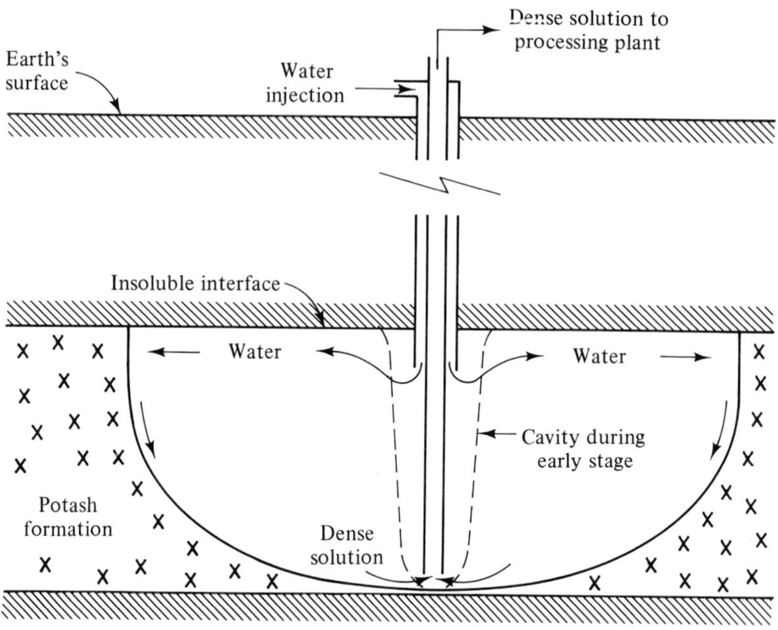

Figure 10.5 Solution mining of potash

standing of the flow pattern within the cavity and its effect on the rate of dissolution of the ore. It also requires that we realize how large these cavities might be. For instance, the depth of the cavity might be of the order of 30 m. The diameter, of course, increases as the ore is removed, and in an old cavity it might be one to two times the depth. At any rate, for most of the time the cavity is in operation it is filled with a large volume of solution whose movement or circulation is probably affected only to a minor extent by the streams entering and leaving the cavity. It would seem reasonable therefore that the entire operation of dissolving the ore is largely controlled by the natural convective pattern of flow that is established as the ore goes into solution. The work of Husband (4) on mass transfer from tall vertical surfaces should be useful for estimating the rate at which the cavity increases in diameter, at least at the top.

Suppose a cavity 10 m in diameter and 10 m high has been washed out of a formation of high-purity KCl. We will attempt to estimate the rate at which KCl can be removed from the upper regions of the cavity using the correlations developed by Husband (4) and given earlier as equations (10.31) and (10.32). We will assume that the upper wall of the cavity is essentially vertical over a height of 3 m and that the boundary layer flow is essentially laminar. Equation (10.31) will therefore be used to estimate the mass-transfer coefficient.

$$\mathrm{Sh} = 1.21(\mathrm{Gr} \times \mathrm{Sc})^{0.26} \tag{10.31}$$

In this equation,

$$\text{Sh} = \frac{k_x H}{D_{AB}}$$

where k_x = mass transfer coefficient defined in terms of mass fractions
H = wall height
D_{AB} = diffusivity of KCl (A) in water (B)

The product of Gr and Sc is the Rayleigh number,

$$\text{Gr} \times \text{Sc} = \text{Ra} = \frac{g\beta_x(x_{A0} - x_{A\infty})H^3}{\nu D_{AB}}$$

where g = acceleration due to gravity, cm/sec^2
β_x = rate of change of mass density with concentration of KCl = $(1/\rho_\infty)(\partial\rho/\partial x_A)$
ρ_∞ = mass density of the solution far from the dissolving wall
$x_{A0} - x_{A\infty}$ = difference in mass fraction of KCl at the dissolving surface and at a large distance from this surface
ν = kinematic viscosity of the boundary layer, cm^2/sec

We will first estimate the Rayleigh number.

$$g = 980 \text{ cm/sec}^2$$

β_x can be determined from density data in the *Handbook of Chemistry and Physics* (17), assuming a cavity temperature of 20°C. Assuming ρ_∞ is the density of pure water at 20°C, we obtain

$$\beta_x = 0.614$$

The solubility of KCl in water at 20°C = 34.0 g/100 g water. Hence

$$x_{A0} - x_{A\infty} = \frac{34.0}{134.0} - 0 = 0.254$$

$$H = 3 \text{ m}$$

$$D_{AB} = 1.27 \text{ cm}^2\text{/day at 10°C} \quad \text{(ref. 17)}$$

$$= 1.27 \times \frac{1}{24 \times 3600} \times \left(\frac{293}{283}\right)^{1.5}$$

$$= 1.55 \times 10^{-5} \text{ cm}^2\text{/sec at 20°C}$$

$$\nu \text{ for water at 20°C} = 100 \text{ cm}^2\text{/sec}$$

The Rayleigh number is

$$\text{Ra} = \frac{980 \times 0.614 \times 0.254(3 \times 100)^3}{100 \times 1.55 \times 10^{-5}} = 2.66 \times 10^{12}$$

From equation (10.31), the calculated Sherwood number is

$$\text{Sh} = 1.21(2.66 \times 10^{12})^{0.26} = 1700$$

and the mass-transfer coefficient is

$$k_x = 1700\left(\frac{D_{AB}}{H}\right) = \frac{1700 \times 1.55 \times 10^{-5}}{3 \times 100} = 8.78 \times 10^{-5} \text{ cm/sec}$$

The production of KCl from the upper part of the cavity can now be estimated. The rate of mass transfer at the wall is

$$N_{A0} = k_x\bar{\rho}\,(x_{A0} - x_{A\infty}) \text{ g/cm}^2\text{-sec}$$

where $\bar{\rho}$ = average density of the boundary layer assumed to be for a 10% solution = 1.06 g/cm^3. Hence

$$N_{A0} = 8.78 \times 10^{-5} \times 1.06 \times 0.254 = 2.36 \times 10^{-5} \text{ g/sec-cm}^2$$

and the production rate is

$$m_A = 2.36 \times 10^{-5}(\pi \times 10 \times 100 \times 3 \times 100) \times \frac{3600}{1000}$$

$$= 80 \text{ kg/hr}$$

10.4 Mass Transfer between Two Fluids

Instead of a fluid in contact with a solid, suppose we now consider two immiscible fluids, designated 1 and 2, in contact with each other. If fluid 1 has dissolved in it a substance s that is also soluble in fluid 2, then as soon as the two fluids are brought together substance s will begin to diffuse into fluid 2. As long as the two phases remain in contact, the transport of s will continue until a condition of equilibrium is established.

The situation imagined here occurs in a variety of ways in the practice of engineering, in processes such as gas absorption, stripping, rectification, humidification, and liquid–liquid extraction. In all these separation processes, two immiscible fluids are brought into contact and one or more components are transferred from one fluid phase to the other. In the engineering design of these processes, it is most important that we understand what is taking place and that we be able to predict the rate at which the mass transfer occurs.

Returning to our system of fluids 1 and 2 with s the transported component, the concentration gradients in the region of the interface between the two

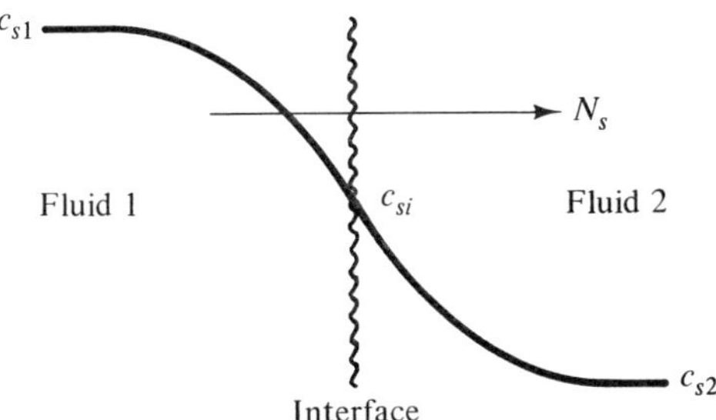

Figure 10.6 Concentration gradients near the interface between immiscible fluids 1 and 2

fluids are as illustrated in Figure 10.6. Concentrations c_{s1} and c_{s2} are the bulk phase concentrations in fluids 1 and 2, respectively, c_{si} is the concentration of s at the interface, and N_s is the molar flux of s. For steady-state conditions, we can define the flux in the same way as in equation (10.4); that is

$$N_s = k_{c1}(c_{s1} - c_{si}) = k_{c2}(c_{si} - c_{s2}) = K_c(c_{s1} - c_{s2}) \tag{10.33}$$

where k_c = individual mass-transfer coefficient defined in terms of the concentration difference in a single phase

K_c = overall mass-transfer coefficient defined in terms of the overall difference in concentration

We notice that the coefficients in equation (10.33) are defined in a way that is analogous to that done in heat transfer, where the individual coefficients h are related to the overall coefficient U (see Section 7.1). This picture of the mass-transfer process is clearly the stagnant-film model again, but with a resistance to transfer on both sides of the interface. For that reason it is often referred to as the two-film model.

From equation (10.33), it readily follows that

$$\frac{1}{k_{c1}} + \frac{1}{k_{c2}} = \frac{1}{K_c} \tag{10.34}$$

indicating that the overall resistance is equal to the sum of the individual resistances in series.

In equation (10.33), the potential for mass transfer is expressed in terms of concentration. However, this is not always the most convenient way to express it. For instance, if fluid 1 is a gas and fluid 2 a liquid, as in gas absorption, the potential in the gas phase is often expressed in terms of partial pressures, while that in the liquid phase may be expressed in terms of concentrations. The expression for the flux is then written as follows for the two individual phases:

$$N_s = k_p(p_{sG} - p_{si}) = k_c(c_{si} - c_{sL}) \tag{10.35}$$

where k_p = individual mass-transfer coefficient for the gas phase with the potential defined in terms of partial pressures

p_{sG}, c_{sL} = partial pressure and concentration of s in the bulk gas and liquid phases, respectively

p_{si}, c_{si} = partial pressure and concentration of s, respectively, at the interface

At the interface, it is usually assumed that the two phases are in equilibrium. This means that p_{si} and c_{si} are related by an equilibrium relationship such as Henry's law if the solutions are dilute:

$$p_{si} = H\, c_{si} \qquad (10.36)$$

where H is Henry's law constant.

The flux N_s can also be expressed in terms of overall mass-transfer coefficients in the following way:

$$N_s = K_p(p_{sG} - p_{sE}) = K_c(c_{sE} - c_{sL}) \qquad (10.37)$$

where K_p = overall mass-transfer coefficient with the overall potential defined in terms of partial pressures

K_c = overall mass-transfer coefficient with the overall potential defined in terms of concentrations

The overall potentials in equation (10.37) are clearly different from those in equation (10.33). The partial pressure potential contains the term p_{sE}, which is a fictitious pressure defined by a gas in equilibrium with the bulk liquid concentration c_{sL}. Thus, if (10.36) is valid,

$$p_{sE} = Hc_{sL} \qquad (10.38)$$

Similarly, the concentration potential in equation (10.37) contains the term c_{sE}, which is a fictitious concentration defined by a liquid in equilibrium with a gas having the bulk gas partial pressure p_{sG}. Hence

$$p_{sG} = Hc_{sE} \qquad (10.39)$$

The relationship between the individual and overall coefficients is readily obtained through the use of equations (10.35) to (10.39). The result is

$$\frac{1}{K_p} = \frac{1}{k_p} + \frac{\mathrm{H}}{k_c} = \frac{\mathrm{H}}{K_c} \qquad (10.40)$$

Partial pressures and concentration are not the only means used to express the potential in mass transfer. For instance, mole fractions are often used, represented by y for a gaseous phase and x for a liquid phase. The steady-state flux of s is then expressed as follows:

$$N_s = k_y(y_{sG} - y_{si}) = k_x(x_{si} - x_{sL})$$

$$= K_y(y_{sG} - y_{sE}) = K_x(x_{sE} - x_{sL}) \qquad (10.41)$$

As before, the subscript E represents a fictitious mole fraction of a phase in equilibrium with the bulk composition in the other phase. The relationship between the individual and overall transfer coefficients is expressed by a relationship similar to equation (10.40):

$$\frac{1}{K_y} = \frac{1}{k_y} + \frac{H}{k_x} = \frac{H}{K_x} \tag{10.42}$$

where H is the constant of proportionality (or Henry's law constant) between y and x for phases in equilibrium; that is,

$$y_{si} = Hx_{si}, \qquad y_{sE} = Hx_{sL}, \qquad y_{sG} = Hx_{sE} \tag{10.43}$$

It should be noted that the two-film model for mass transfer assumes that all the resistance to mass transfer lies in the two "films" that lie on either side of the interface, and that there is no resistance to mass transfer at the interface itself. This may not be a valid assumption in many cases. For instance, if the system contains only a very small amount of surface active substance, the rate of mass transfer can be markedly reduced, because such substances congregate at the interface between phases and form an additional resistance to mass transfer. Other phenomena, such as interfacial turbulence or chemical reaction near the phase boundary, can markedly affect the rate of mass transfer as well. If such processes are not present when the coefficients are measured and the data correlated, the predicted and actual rates of mass transfer will be quite different.

10.4.1 Measurement of the Mass Transfer Coefficient k_c. Equation (10.33) defines the coefficient k_c as the ratio of the flux and the concentration potential.

$$k_c = \frac{N_s}{\Delta c_s} = \frac{m_s}{A\ \Delta c_s} \tag{10.44}$$

The determination of k_c therefore requires a system where the three quantities m_s, A, and Δc_s can be accurately measured.

The interfacial area A is well defined for single bubbles of gas and for single droplets of liquid. It is not surprising therefore that the literature contains many reports of studies on mass transfer carried out in this way. The results of these investigations have been recently reviewed (10) and will not be discussed here. A third technique that has also been widely used is to have a vertical tube with a layer of liquid flowing downward over the inner surface and a stream of gas passing in either the upward or downward direction. Such a column, illustrated in Figure 10.7, has a well-defined interfacial area for mass transfer provided there are no ripples on the surface of the liquid. If the liquid is pure and vaporizes into the gas stream, a concentration potential exists only in the gas stream, and the system can be used to measure k_c for the gas phase. By changing flow rates, the tube diameter, system temperature, and pressure, and by using different liquids and gases, mass

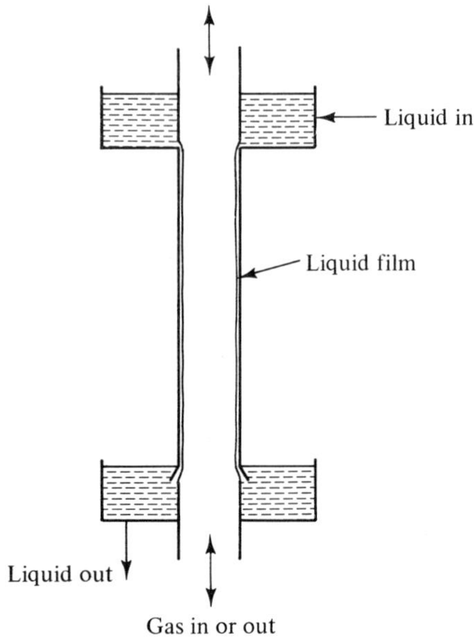

Figure 10.7 Wetted-wall column

transfer can be studied over a wide range of conditions. The following calculation, based on the early work of Gilliland and Sherwood (2) is an illustration of how k_c is determined from measurements made with such a column.

Example 10.4.1

The column used by Gilliland and Sherwood (2) had a wetted section that was 117 cm long with an inner diameter of 2.67 cm. In run Wlp water flowed down the inner wall of this pipe and a stream of air passed in the same direction. The experiment thus serves to determine k_c for water vapor being transferred into the stream of air.

The measured rate of water vaporization was 3.8 cm³/min, which corresponds to

$$m_s = 3.8 \times 1.0 \times 60/18 = 13 \text{ g mol/hr}$$

The transfer area is

$$A = \frac{\pi \times 2.67 \times 117}{100 \times 100} = 0.0981 \text{ m}^2$$

The potential Δc_s is determined indirectly from several measurements. From the measured water temperature at the top and bottom of the column, the vapor pressure of the water at these points is determined from available data. Thus

$$T \text{ (top)} = 31.1^\circ\text{C} \qquad p_w^\circ \text{ (top)} = 33.9 \text{ mm Hg}$$

$$T \text{ (bottom)} = 26.1^\circ\text{C}, \qquad p_w^\circ \text{ (bottom)} = 25.4 \text{ mm Hg}$$

The wet-bulb temperature of the air entering and leaving the column was measured and used to determine the absolute humidity of the air at these points.

$$H \text{ (top)} = 0.002 \text{ g water/g dry air}$$

$$H \text{ (bottom)} = 0.016 \text{ g water/g dry air}$$

The relationship between humidity H and partial pressure p_w of water vapor is

$$H = \frac{M_w p_w}{M_a(p_t - p_w)}$$

where M_w, M_a = molecular weight of water and air, respectively, and p_t = total pressure. Solving for p_w,

$$p_w = \frac{HM_a p_t}{M_w + HM_a}$$

The measured total pressure was $p_t = 770$ mm Hg. Hence

$$p_w \text{ (top)} = \frac{0.002 \times 28.9 \times 770}{18 + 0.002 \times 28.9} = 2.5 \text{ mm Hg}$$

$$p_w \text{ (bottom)} = \frac{0.016 \times 28.9 \times 770}{18 + 0.016 \times 28.9} = 19.3 \text{ mm Hg}$$

The potential for mass transfer is thus

$$\Delta p_w \text{ (top)} = 33.9 - 2.5 = 31.4 \text{ mm Hg}$$

$$\Delta p_w \text{ (bottom)} = 25.4 - 19.3 = 6.1 \text{ mm Hg}$$

This may be converted to concentration units using the ideal gas law and the measured temperatures of the air stream.

$$\Delta c_w \text{ (top)} = \frac{\Delta p_w}{RT} = \frac{31.4}{760 \times 0.08205 \times 304.0} = 1.66 \times 10^{-3} \text{ g mol/liter}$$

$$\Delta c_w \text{ (bottom)} = \frac{6.1}{760 \times 0.08205 \times 300.2} = 0.326 \times 10^{-3} \text{ g mol/liter}$$

The mean value of Δc_w over the length of the column is taken as the logarithmic mean. Thus, for use in equation (10.44),

$$\Delta c_s = \frac{(1.66 - 0.326)10^{-3}}{\log_e (1.66/0.326)} = 0.82 \times 10^{-3} \text{ g mol/liter}$$

The value of the coefficient k_c is, therefore,

$$k_c = 13 \frac{\text{g mol}}{\text{hr}} \times \frac{1}{0.0981 \text{ m}^2} \times \frac{\text{liter}}{0.82 \times 10^{-3} \text{ g mol}} \times 10^{-3} \frac{\text{m}^3}{\text{liter}}$$

$$= 1.6 \times 10^2 \text{ m/hr}.$$

10.4.2 Correlation of Data Obtained with Wetted-Wall Columns. The report by Gilliland and Sherwood (2) contains the data from many experiments with water and eight different organic liquids, all vaporizing into a stream of air. Correlation of all these data was accomplished by an expression similar in form to equation (10.19). However, instead of the Sherwood number, the authors used the dimensionless parameter D_t/x, where D_t is the inside diameter of the wetted-wall column, and x is a film thickness defined by the Stefan diffusion equation (9.21).

$$x = \frac{D_{AB}p_t}{RTN_A}\frac{\Delta p_A}{p_{BM}} \tag{9.21}$$

where D_{AB} = diffusion coefficient of the vapor in air

N_A = steady-state flux of vapor A

Δp_A = potential for mass transfer of A expressed in partial pressures

p_{BM} = logarithmic mean partial pressure of the air over the length of column

Using equations (10.44) and (9.21), as well as the ideal gas law in the form

$$\Delta c_A = \frac{\Delta p_A}{RT}$$

the ratio D_t/x can be written as

$$\frac{D_t}{x} \equiv \frac{k_c D_t}{D_{AB}}\frac{p_{BM}}{p_t} \equiv \text{Sh}\,\frac{p_{BM}}{p} \tag{10.45}$$

D_t/x can thus be considered as a modified Sherwood number. The correlation obtained by the authors (2) is expressed by the equation

$$\frac{k_c D_t}{D_{AB}}\frac{p_{BM}}{p_t} = 0.023\,\text{Re}^{0.83}\,\text{Sc}^{0.44} \tag{10.46}$$

where the Reynolds number is based on the tube diameter and the velocity of the air relative to the tube.

Since the early work of Gilliland and Sherwood, many studies have been carried out using the wetted-wall column. The result has been to modify the constants but not the form of equation (10.46). For instance, in a recent review of this work [page 214 of reference (10)], it is suggested that the following correlation of Johnstone and Pigford be used:

$$\frac{k_c D_t}{D_{AB}}\frac{p_{BM}}{p_t} = 0.0328\,(\text{Re}')^{0.77}\,(\text{Sc})^{0.33} \tag{10.47}$$

where Re′ is the Reynolds number based on the gas velocity relative to the velocity of the liquid surface. If the liquid surface velocity cannot be determined, these authors still recommend that equation (10.46) be used to obtain the approximate value of k_c.

10.4.3 Mass Transfer Coefficients for Packed Columns.

A system widely used in industry for mass transfer between gas and liquid phases is the packed column. This is essentially a vertical shell or column filled with small pieces of solid material. Liquid is distributed over the packing at the top of the column to flow downward under the influence of gravity over the surface of the packing and be removed at the bottom of the column. Gas is introduced beneath the bottom layer of packing, flows upward through the spaces between the pieces of packing, and is removed at the top of the column. In such a system the interfacial area for mass transfer is a complex function of the shape and size of the packing, as well as of the flow rates of the two fluid streams. Unlike the situation with the wetted-wall column, this area is difficult to measure in the packed column and is usually unknown.

A new variable is therefore introduced into the calculations, represented by the symbol a and defined as the interfacial area for mass transfer per unit volume of space occupied by the packing. If equation (10.33) is multiplied by this variable, we obtain

$$N_s a = k_{c1} a (c_{s1} - c_{si}) = k_{c2} a (c_{si} - c_{s2}) = K'_c a (c_{s1} - c_{s2}) \qquad (10.48)$$

from which

$$\frac{1}{k_{c1} a} + \frac{1}{k_{c2} a} = \frac{1}{K'_c a} \qquad (10.49)$$

The product $N_s a$ is the rate of mass transfer of s per unit volume of column with the units mol hr^{-1}m^{-3}. The terms $k_c a$ and $K'_c a$ are mass-transfer coefficients or capacity coefficients on a volume basis with units of reciprocal time, hr^{-1}. Similar coefficients are obtained when the potential is expressed in partial pressures or in mole fractions, such as $k_p a$, $K_p a$, $k_y a$, $k_x a$, $K_y a$, and $K_x a$.

Numerical values for the capacity coefficients are obtained empirically from performance tests. The literature on mass transfer contains several reports of this kind, a good example of which is the work of Whitney and Vivian (15) on the absorption of SO_2 from air by water.

10.4.4 Capacity Coefficients for the Absorption of SO_2 in Water.

The column used by Whitney and Vivian (15) was constructed of lead and had an internal diameter of 20.3 cm and a depth of packing of about 30 cm (the authors' data have been converted to SI or cgs units). The packing was standard ceramic Raschig rings of size 25.4 mm. Gas from the combustion of sulfur was mixed with air, passed through a water scrubber to remove SO_3 and to saturate the gas with water vapor, and then passed to the bottom of the absorption column. Plant water was sent to a constant-head tank, from which it flowed at a controlled rate to the distributor at the top of the packing. With flows and temperature constant, the absorber was allowed to reach a steady-state condition, at which point the gas stream entering and leaving the absorber and the exit liquid stream were sampled and subsequently analyzed.

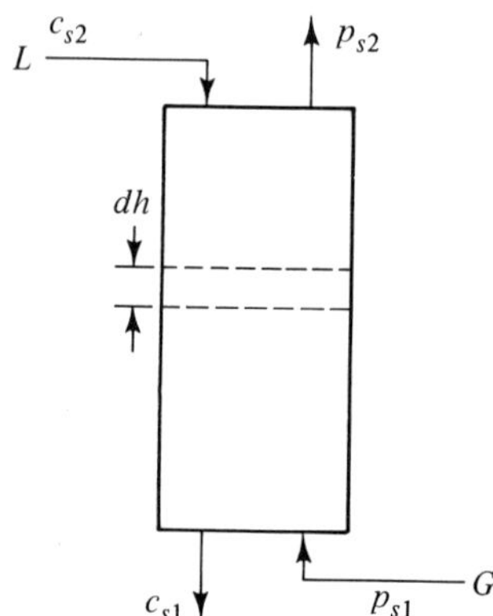

Figure 10.8 Countercurrent absorber

The method of calculation used to obtain the capacity coefficient will be given in detail for one of the runs reported by Whitney and Vivian. The flux of SO_2 at any point in the absorber is written in the following way, which is similar to equation (10.37):

$$N_s a = K_L a (c_{sE} - c_{sL}) \tag{10.50}$$

where $N_s a$ = moles SO_2 transferred from the gas to the liquid phase/hr/m^3 of absorber volume

$K_L a$ = overall capacity coefficient based on liquid phase concentration, hr^{-1}

c_{sE} = fictitious liquid phase concentration in equilibrium with the partial pressure of SO_2 in the bulk gas phase, mol/m^3

c_{sL} = concentration of SO_2 in the bulk liquid phase, mol/m^3

For an absorber of unit cross-sectional area (Figure 10.8), the flux of SO_2 at the differential section of height dh is also given by the increase of SO_2 in the liquid phase:

$$N_s a = \frac{L}{\rho} \frac{dc}{dh} \tag{10.51}$$

where L = liquid flow rate in kg/hr/m^2 of column cross section

ρ = liquid density, kg/m^3

Equating equations (10.50) and (10.51) and rearranging the result in integral form gives

$$K_L a \int_0^h dh = \frac{L}{\rho} \int_{c2}^{c1} \frac{dc}{c_{sE} - c_{sL}} \tag{10.52}$$

where it is assumed that L, ρ, *and* $K_L a$ are constant over the height h of the absorber. Solving for the coefficient $K_L a$,

$$K_L a = \frac{L}{\rho h} \int_{c2}^{c1} \frac{dc}{c_{sE} - c_{sL}} \tag{10.53}$$

The calculations will be illustrated for run 34.

$$L = 920 \text{ lb/hr-ft}^2 = 4492 \text{ kg/hr-m}^2$$

$$h = 24.3 \text{ in} = 0.617 \text{ m}$$

$$\rho \text{ at } 67°\text{F} = 0.998 \text{ g/cm}^3 = 998 \text{ kg/m}^3$$

$$c_1 = 0.00447 \text{ lb-mol/ft}^3 = 0.0716 \text{ kg-mol/m}^3$$

$$c_2 = 0$$

$$p_{s1} = 0.060 \text{ atm}$$

$$p_{s2} = 0.032 \text{ atm}$$

Since the change in concentration over the height of the absorber is not large, a linear variation in Δc_s can be used. Thus, the integral in equation (10.53) may be approximated as

$$K_L a = \frac{L}{\rho h} \frac{c_1 - c_2}{(c_{sE} - c_{sL})_{1m}} \tag{10.54}$$

where $(c_{sE} - c_{sL})_{1m}$ = the logarithmic mean concentration difference at the bottom and top of the absorber. From data in the report (15),

$$c_{sE1} = 0.0080 \text{ lb-mol/ft}^3 = 0.128 \text{ kg-mol/m}^3$$

$$c_{sE2} = 0.0049 \text{ lb-mol/ft}^3 = 0.0785 \text{ kg-mol/m}^3$$

Hence

$$(c_{sE} - c_{sL})_{1m} = \frac{(0.128 - 0.0716) - (0.0785 - 0)}{\log_e (0.0564/0.0785)} = 0.0668 \text{ kg-mol/m}^3$$

and

$$K_L a = \frac{4492}{998 \times 0.617} \times \frac{0.0716 - 0}{0.0668} = 7.8 \text{ hr}^{-1}$$

Whitney and Vivian (15) made similar calculations for a large number of runs in which the effect of the following variables was studied: water flow rate, gas flow rate, and temperature. At each temperature studied, they were able to separate the overall coefficients into the two individual coefficients and to express these in terms of the water and gas flow rates. For instance, at 21.1°C (70°F) the correlations obtained were expressed as follows:

$$k_L a = 0.16 L^{0.82} \tag{10.55}$$

$$k_G a = 8.0 \times 10^2 G^{0.7} L^{0.25} \tag{10.56}$$

where k_La = liquid film mass-transfer coefficient, hr^{-1}

k_Ga = gas film mass-transfer coefficient, kg-mol hr^{-1} m^{-3} Pa^{-1}

L = liquid rate per unit of tower cross-sectional area, kg hr^{-1} m^{-2}

G = average total gas rate, at top and bottom of tower per unit of tower cross-sectional area, kg hr^{-1} m^{-2}

It should be remembered that these values for the coefficients are specific for the type of packing used in the study, that is, 25.4-mm ceramic Raschig rings.

10.5 Estimate of Tower Size Required to Remove SO_2 from Air

Suppose the SO_2 in a stream of air is to be removed by scrubbing with water. We can use the data of Whitney and Vivian (15) to estimate the size of a packed tower that is required for this operation.

Assume we have a stream of gas measuring 10,000 m^3/hr at 21.1°C (70°F) and 1.2 atm, saturated with water vapor and containing 5% SO_2. If this gas is to be scrubbed with water to remove 95% of the SO_2, estimate the required tower diameter and depth of packing if 25.4-mm ceramic rings are used and the tower is operated essentially at a constant temperature of 21.1°C.

Suppose we choose a water flow rate of 34,200 kg/hr-m^2 (7000 lb/hr-ft^2) in the range used by Whitney and Vivian (15). Using Figure 8 of reference 15, the gas flow rate and overall coefficient can be estimated. The gas rate is taken at about one-half the flow at the flooding velocity.

$$G = 2100 \text{ kg/hr-m}^2 \text{ (430 lb/hr-ft}^2\text{)}$$

$$K_La = 45 \text{ hr}^{-1}$$

The mass-flow rate of the available gas stream is, using the ideal gas equation,

$$m = \frac{pVM}{RT} = \frac{1.2 \times 1.013 \times 10^5 \times 10{,}000 \times 29}{8314 \times 294.1} = 14{,}420 \text{ kg/hr}$$

where M = molecular weight of the gas stream, assumed to be 29. The ratio m/G gives the tower cross-sectional area:

$$\text{Area} = m/G = \frac{14{,}420}{2100} = 6.87 \text{ m}^2$$

from which the tower diameter is obtained.

$$D = \text{tower diameter} = \left(\frac{4 \times 6.87}{\pi}\right)^{1/2} = 2.96 \approx 3.0 \text{ m}$$

The total water flow rate is now known:

$$\text{Water flow} = 34{,}200 \times 6.87 = 235{,}000 \text{ kg/hr}$$

An overall material balance for the SO_2 can be used to determine the terminal concentrations c_1 and p_2 (where p is the partial pressure of SO_2). Thus the SO_2 removed from the gas stream is

$$\frac{0.95 \times 0.05 \times 1.2 \times 1.013 \times 10^5 \times 10{,}000}{8314 \times 294.1} = 23.6 \text{ kg-mol/hr}$$

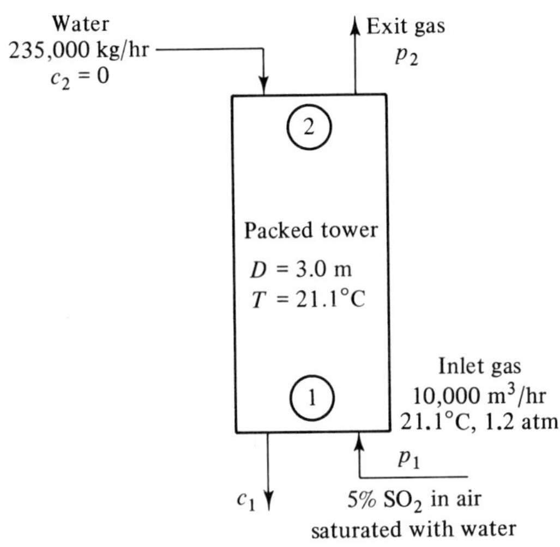

The density of water at 21.1°C = 0.998 g/cm^3 = 998 kg/m^3. Since c_2 is zero, the value of c_1 is

$$c_1 = \frac{23.6 \times 998}{235{,}000} = 0.100 \text{ kg-mol/m}^3 = 0.0063 \text{ lb-mol/ft}^3$$

The value of p_2 is obtained as follows. The amount of SO_2 in the exit gas is

$$\frac{0.05 \times 0.05 \times 10{,}000 \times 1.2}{1.0} = 30.0 \text{ m}^3\text{/hr}$$

at an assumed exit pressure of 1.0 atm. The inerts in the exit gas stream amount to

$$\frac{0.95 \times 10{,}000 \times 1.2}{1.0} = 11{,}400 \text{ m}^3\text{/hr}$$

Hence the partial pressure of SO_2 in the exit gas is

$$p_2 = \frac{30.0 \times 1.0}{11400 + 30.0} = 0.0026 \text{ atm}$$

To find the required depth of packing, equation (10.54) is solved for h.

$$h = \frac{L}{\rho K_L a} \frac{c_1 - c_2}{(c_{sE} - c_{sL})_{1m}} \tag{10.54}$$

This assumes that the equilibrium curve is linear and that the concentration of SO_2 also changes linearly over the length of the column. Both assumptions are probably not unreasonable in this case. Using the reported equilibrium data (15),

$$c_{sE1} = 0.0080 \text{ lb-mol/ft}^3 = 0.128 \text{ kg-mol/m}^3$$

$$c_{sE2} \approx 0.0010 \text{ lb-mol/ft}^3 = 0.016 \text{ kg-mol/m}^3$$

and

$$(c_{sE} - c_{sL})_{1m} = \frac{(0.128 - 0.100) - (0.016 - 0)}{\log_e (0.028/0.016)} = 0.021 \text{ kg-mol/m}^3$$

From equation (10.54),

$$h = \frac{34{,}200}{998 \times 45} \times \frac{0.100 - 0}{0.021} = 3.6 \text{ m}$$

Most towers have a larger ratio of packing depth to tower diameter. We can increase this ratio by reducing the flow of water. However, there is a minimum water flow below which we cannot operate the tower. We can also reduce the tower diameter by operating with a higher gas flow rate G. But this also must be kept below the loading point, which is the value of G where the gas flow begins to impede the downward flow of the liquid.

REFERENCES

1. Danckwerts, P. V., *Ind. Eng. Chem.*, *43*, 1460 (1951).
2. Gilliland, E. R., and T. K. Sherwood, *Ind. Eng. Chem.*, *26*, 516 (1934).
3. Higbie, R., *Trans AI ChE*, *31*, 365 (1935).
4. Husband, W. H., Ph.D. thesis, University of Saskatchewan (1969). University Microfilms, Ann Arbor, Mich., order number 69-16743.
5. King, C. J., *Separation Processes*. New York: McGraw-Hill Book Company, 1980, p. 520.
6. Schenkels, F. A. M., and J. Schenk, *Chem. Eng. Sci.*, *24*(3), 585–93 (1969).
7. Schuepp, P. H., *J. Appl. Meteorol.*, *10*(5), 1018–25 (1971).
8. Schütz, G., *Int. J. Heat Mass Transfer*, *6*, 873–79 (1963).
9. Sherwood, T. K., and R. L. Pigford, *Absorption and Extraction*. New York: McGraw-Hill Book Company, 1952, p. 66.
10. Sherwood, T. K., R. L. Pigford, and C. R. Wilke, *Mass Transfer*. New York: McGraw-Hill Book Company, 1975, p. 201.
11. Ibid, p. 214.
12. Brian, P. T. L., and H. B. Hales, *AI ChEJ*, *15*, 419 (1969).

13. Sissom, L. E., and D. R. Pitts, *Elements of Transport Phenomena*. New York: McGraw-Hill Book Company, 1972, p. 537.
14. Treybal, R. E., *Mass Transfer Operations*. New York: McGraw-Hill Book Company, 1980, p. 60.
15. Whitney, Roy P., and J. E. Vivian, *Chem. Eng. Progress*, *45*(5), 323–37 (1949).
16. Wilke, C. R., C. W. Tobias, and M. Eisenberg, *Chem. Eng. Progress*, *49*(12), 663–74 (1953).
17. Hodgman, C. D., ed., *Handbook of Chemistry and Physics*, 30th ed. Cleveland: Chemical Rubber Publishing Co., 1946.

PROBLEMS

10.1. **(a)** For Example 10.1.1, calculate the value of k_x.
(b) Calculate the value of k_y for Example 10.1.2.
Answer: (a) 5.8×10^{-4} g mol/cm²-sec; (b) 7.7×10^{3} g mol/m²-hr

10.2. Air at 45°C and with a relative humidity of 20% flows over a smooth water surface at an average velocity of 2 m/s. The length of the water surface in the direction of airflow is 3 m. The average surface temperature of the water is estimated to be 32°C. Calculate the water evaporated per hour per square meter of surface area. Vapor pressure of water at 32°C and 45°C is 35.7 mm Hg and 71.9 mm Hg, respectively. *Answer:* 0.60 kg/m²-hr

10.3. Dry air at 1 atm pressure and 40°C enters a tube that is 10 m long and 15 cm inside diameter at an average velocity of 15 cm/s. The inner surface of the tube is lined with a thin layer of felt material that is kept saturated with water at 16°C. Assuming constant temperature of the air and pipe wall, estimate the amount of water vapor carried out of the tube by the air flow per hour. ν for air at 40°C and 1 atm is 1.85×10^{-5} m²/s. Vapor pressure of water at 16°C is 13.6 mm Hg. *Answer:* 0.164 kg/hr

10.4. A porous metal sphere, 5 cm in diameter and saturated with water, is suspended in a stream of dry air that has a free-stream velocity of 1 m/s. Pressure is 1 atm and the temperature of the air and sphere is 20°C. Estimate the rate of mass transfer from the sphere. ν for air at 20°C and 1 atm is 1.53×10^{-5} m²/s. Vapor pressure of water at 20°C is 17.5 mm Hg. *Answer:* 3.23×10^{-3} kg/hr

10.5. A sphere of naphthalene ($C_{10}H_8$, molecular weight 128), 2.5 cm in diameter, is suspended in a stream of air moving with a free-stream velocity of 1 m/s. Temperature of the air and sphere is 20°C; pressure is 1 atm. Estimate the time required for the diameter of the sphere to be reduced to one-half its initial value. ν for air at 20°C, 1 atm is 1.53×10^{-5} m²/s. Vapor pressure of naphthalene at 20°C is 0.040 mm Hg. Density of solid naphthalene is 1145 kg/m³. *Note:* Under these conditions solid naphthalene sublimes, that is, passes directly from the solid to the vapor phase. *Answer:* 60.5 days

10.6. For conditions as given in Problem 10.5, repeat the calculation of the time assuming mass transfer occurs by natural convection rather than by forced convection. *Answer:* 241 days

Appendix

TABLE A.1
PROPERTIES OF LIQUID WATER

°C	°F	ρ (kg/m^3)	C_p (J/kg-K)	k (W/m-K)	ν (m^2/s)
0	32	999.8	4205	0.5750	1.753×10^{-6}
7	44.6	999.9	4196	0.5818	1.422
27	80.6	996.6	4177	0.6084	0.826
47	116.6	989.3	4177	0.6367	0.566
67	152.6	979.5	4187	0.6587	0.420
87	188.6	967.4	4206	0.6743	0.330
100	212	957.2	4219	0.6811	0.290
127	260.6	937.5	4241	0.6864	0.229×10^{-6}
147	296.6	919.9	4306	0.6836	2.000×10^{-7}
167	332.6	900.5	4391	0.6774	1.786
187	368.6	879.5	4456	0.6672	1.626
207	404.6	856.6	4534	0.6530	1.504×10^{-7}

Adapted with permission from John H. Lienhard, *A Heat Transfer Textbook*, Prentice-Hall, Inc., Englewood Cliffs, New Jersey (1981).

TABLE A.2
PROPERTIES OF AIR AT 1 ATMOSPHERE PRESSURE

T (K)	C_p (J/kg-K)	μ (kg/m-s)	k (W/m-K)
300	1003	1.853×10^{-5}	0.02614
350	1008	2.081	0.02970
400	1013	2.294	0.03305
500	1029	2.682	0.03951
600	1051	3.030	0.0456
700	1075	3.349×10^{-5}	0.0513

Adapted with permission from John H. Lienhard, *A Heat Transfer Textbook*, Prentice-Hall, Inc., Englewood Cliffs, New Jersey (1981).

TABLE A.3
THERMAL CONDUCTIVITY OF SOME SOLID MATERIALS

	k W/m-K	
	100°C	300°C
Aluminum	206	230
Copper	377	367
Steel	45	43
Asbestos (insulating)	0.17	—
Glass wood, fiber glass	0.074	—
Kaolin brick	—	0.26
Kaolin firebrick	—	0.17
Magnesia brick	—	0.064
Fireclay brick	—	1.0
Masonry brick	0.66	—
Glass	0.78	—
85% magnesia	—	0.064

TABLE A.4
PLANCK'S RADIATION FUNCTIONS

λT		$\frac{E_{b(0-\lambda T)}}{\sigma T^4}$	λT		$\frac{E_{b(0-\lambda T)}}{\sigma T^4}$
μm °R	μm K		μm °R	μm K	
1,000	555.6	0	10,400	5,777.8	0.7181
1,200	666.7	0	10,600	5,888.9	0.7282
1,400	777.8	0	10,800	6,000.0	0.7378
1,600	888.9	0.0001	11,000	6,111.1	0.7474
1,800	1,000.0	0.0003	11,200	6,222.2	0.7559
2,000	1,111.1	0.0009	11,400	6,333.3	0.7643
2,200	1,222.2	0.0025	11,600	6,444.4	0.7724
2,400	1,333.3	0.0053	11,800	6,555.6	0.7802
2,600	1,444.4	0.0098	12,000	6,666.7	0.7876
2,800	1,555.6	0.0164	12,200	6,777.8	0.7947
3,000	1,666.7	0.0254	12,400	6,888.9	0.8015
3,200	1,777.8	0.0368	12,600	7,000.0	0.8081
3,400	1,888.9	0.0506	12,800	7,111.1	0.8144
3,600	2,000.0	0.0667	13,000	7,222.2	0.8204
3,800	2,111.1	0.0850	13,200	7,333.3	0.8262
4,000	2,222.2	0.1051	13,400	7,444.4	0.8317
4,200	2,333.3	0.1267	13,600	7,555.6	0.8370
4,400	2,444.4	0.1496	13,800	7,666.7	0.8421
4,600	2,555.6	0.1734	14,000	7,777.8	0.8470
4,800	2,666.7	0.1979	14,200	7,888.9	0.8517
5,000	2,777.8	0.2229	14,400	8,000.0	0.8563
5,200	2,888.9	0.2481	14,600	8,111.1	0.8606
5,400	3,000.0	0.2733	14,800	8,222.2	0.8648
5,600	3,111.1	0.2983	15,000	8,333.3	0.8688
5,800	3,222.2	0.3230	16,000	8,888.9	0.8868
6,000	3,333.3	0.3474	17,000	9,444.4	0.9017
6,200	3,444.4	0.3712	18,000	10,000.0	0.9142
6,400	3,555.6	0.3945	19,000	10,555.6	0.9247
6,600	3,666.7	0.4171	20,000	11,111.1	0.9335
6,800	3,777.8	0.4391	21,000	11,666.7	0.9411
7,000	3,888.9	0.4604	22,000	12,222.2	0.9475
7,200	4,000.0	0.4809	23,000	12,777.8	0.9531
7,400	4,111.1	0.5007	24,000	13,333.3	0.9589
7,600	4,222.2	0.5199	25,000	13,888.9	0.9621
7,800	4,333.3	0.5381	26,000	14,444.4	0.9657
8,000	4,444.4	0.5558	27,000	15,000.0	0.9689
8,200	4,555.6	0.5727	28,000	15,555.6	0.9718
8,400	4,666.7	0.5890	29,000	16,111.1	0.9742
8,600	4,777.8	0.6045	30,000	16,666.7	0.9765
8,800	4,888.9	0.6195	40,000	22,222.2	0.9881
9,000	5,000.0	0.6337	50,000	27,777.8	0.9941
9,200	5,111.1	0.6474	60,000	33,333.3	0.9963
9,400	5,222.2	0.6606	70,000	38,888.9	0.9981
9,600	5,333.3	0.6731	80,000	44,444.4	0.9987
9,800	5,444.4	0.6851	90,000	50,000.0	0.9990
10,000	5,555.6	0.6966	100,000	55,555.6	0.9992
10,200	5,666.7	0.7076	∞	∞	1.0000

From R. V. Dunkle, *Trans. ASME 76*, 550 (1954).

Radiation Shape Factors

$$\text{Let } x = 1 + \left[1 + \left(\frac{r_2}{d}\right)^2\right]\left(\frac{d}{r_1}\right)^2$$

$$F_{A_1 \to A_2} = \frac{1}{2}\left[x - \sqrt{x^2 - 4\left(\frac{r_2}{d}\right)^2\left(\frac{d}{r_1}\right)^2}\right]$$

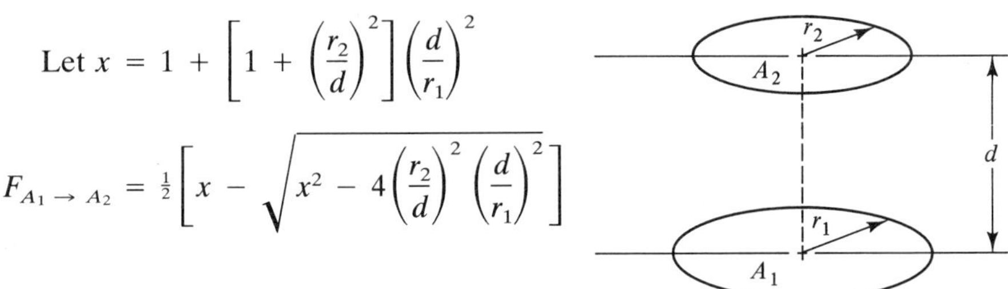

Figure A.1 Geometric factor for parallel disks

Figure A.2 Geometric factor for parallel rectangles (Figures A.2 and A.3 from Schenck, Jr., Hilbert, *Heat Transfer Engineering*, © 1959, p. 133. Adapted by permission of Prentice-Hall, Englewood Cliffs, New Jersey)

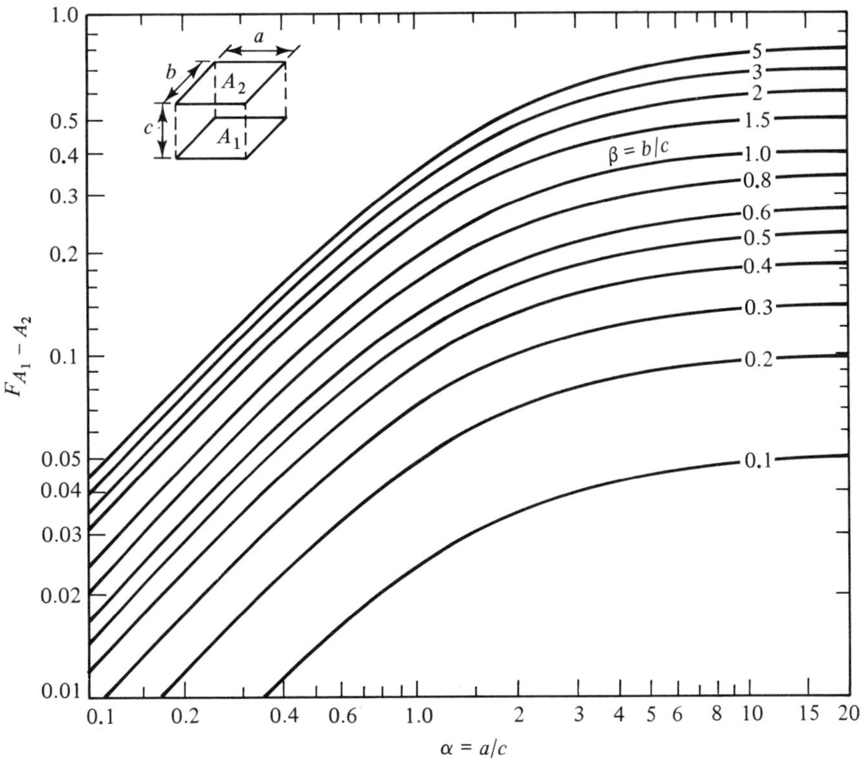

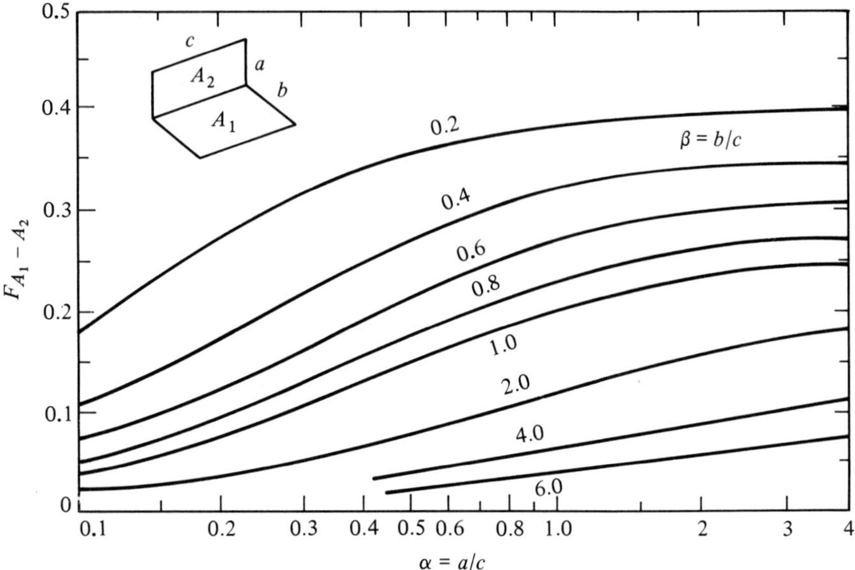

Figure A.3 Geometric factor for rectangles at right angles with a common edge

TABLE A.5
MOLECULAR WEIGHT OF SOME SUBSTANCES

Substance	Chemical Formula	Molecular Weight (M)
Air	—	28.9
Argon	Ar	39.95
Benzene	C_6H_6	78.11
Carbon	C	12.00
Carbon dioxide	CO_2	44.01
Carbon monoxide	CO	28.01
Chlorine	Cl_2	70.91
Hydrogen	H_2	2.016
Iodine	I_2	253.81
Methane	CH_4	16.04
Neon	Ne	20.18
Nitrogen	N_2	28.01
Oxygen	O_2	32.00
Sodium	Na	23.00
Sulfur	S	32.06
Water	H_2O	18.02

TABLE A.6
BINARY DIFFUSION COEFFICIENTS AT 1 ATMOSPHERE PRESSURE

A/B	T (°C)	D_{AB} (cm^2/hr)
Acetone/air	0	393
Benzene/air	0	278
Carbon dioxide/nitrogen	20	587
Oxygen/nitrogen	0	652
Water vapor/air	0	792
Water vapor/air	20	881

Index

Absorption:
  of SO_2 in water, 247
  tower size to remove SO_2 from air, 250
Absorptivity, for radiation, 171
Air, properties, 255

Bernoulli's equation, 26
Black body:
  Planck's equation for monochromatic emissive power, 174
  radiant heat exchange, 189
Black-body radiation, 172
Blasius, 88, 97
Boundary layer equations:
  energy, 85
  mass conservation, 82
  momentum, 83
Boundary layer flow, 70, 76
  displacement thickness δ_1, 77
  kinetic energy thickness δ_3, 78
  momentum thickness δ_2, 78
Boundary layer growth:
  flat plate, laminar, 91
  flat plate, turbulent, 98
Boundary layer shape, turbulent:
  shear stress, wall, 95
Boundary layer velocity profile:
  polynomial, 93
  sine curve, 80, 91
  turbulent, 95

Coefficient of heat transfer (*see* Heat transfer coefficient)
Conduction of heat (*see* Heat conduction)
Configuration factor, radiation, 177, 180
  parallel disks, 257
  parallel planes, 257
  planes at right angles, 258
  relationships, 181
  special reciprocity relation, 186
Continuity, equation of, 24
Convective heat transfer, 133
Critical radius (*see* Insulated pipes)

Diffusion, molecular:
  catalytic surface, oxidation of SO_2 in an infinite stagnant medium, 219
  coal particle, combustion in an infinite stagnant medium, 217
  convective flow during, 209
  counterdiffusion, equimolar, 213
  Fick's equation, 12, 205
  spherical droplet, evaporation into an infinite stagnant medium, 215
  through a stagnant gas, 210
  Stefan's experiment, 208
Diffusion coefficient:
  experimental values for gases, 259
  Gilliland correlation, 206
  organic substances in liquids, Wilke correlation, 207
  water vapor in air, Spalding correlation, 207
Diffusivity:
  definition, 12
  gases, 259
Dimensional analysis:
  for forced convection, 144
  mass transfer to a plane surface, 228
  pipe friction, 47
Drag force on a submerged body, 66

Electrical analog for radiating systems, 192
Emissive power, radiation, 174
Emissivity, definition, 173
Energy equation, steady flow, 44
Euler's equation, 26

Fick's equation (*see* Diffusion, molecular)
Fins:
- annular, 127
- efficiency, 126
- heat transfer from, 120, 124
- rectangular, 126
- triangular, 126

Flow description:
- Bernoulli's equation, 26
- dimensional analysis and pipe friction, 47
- energy equation along a steamline, 25
- energy equation with friction, 44
- Euler's equation, 26
- kinetic energy correction factor α, 46
- laminar and turbulent, 22
- laminar flow in pipes, 55
- mathematical, 22
- minor losses in pipe flow, 58
- momentum equation, linear, 27
- one-, two-, and three-dimensional, 22
- steady and unsteady, 21
- streamlines, 22

Flow measurement:
- nozzle, 32
- orifice meter, 34
- Pitot tube, 31
- venturi meter, 32

Fourier's equation (*see* Heat conduction)
Friction factor, pipe flow, 50

Gilliland correlation (*see* Diffusion coefficient)
Gray surface, radiation, 190

Hagen-Poiseuille equation, for laminar flow in a tube, 56
Heat conduction:
- composite cylinder, 110
- cylinder with heat generation, 118
- cylindrical tube, 108
- fins, long rod, 122
- fins with insulated tip, 122
- Fourier's equation of, 10, 103
- insulated pipes, 112
- sphere, thick-walled, 111
- wall, composite, 105
- wall, plane, 104
- wall with heat generation, 113
- wall with heat generation and convection, 117

Heat exchangers:
- double-pipe, 157
- mean temperature difference in a double-pipe exchanger, 160
- mean temperature difference in multipass exchangers, 162
- multipass, 162
- overall coefficient of heat transfer, 159

Heat transfer by radiation (*see* Radiant heat transfer)
Heat transfer coefficient:
- definition, 106, 133
- determination by cooling sphere, 134
- determination by optical interferometer, 135
- determination using boundary layer analysis, 137
- for flow over flat plates, 149
- for flow over spheres, 148
- forced convection correlation, 144
- for fouling deposits, 167
- for gases flowing across a cylinder, 148
- laminar flow in pipes correlation, 147
- natural convection correlation, 146
- natural convection from horizontal cylinders, 149
- natural convection from spheres, 149
- natural convection for vertical planes, 150
- overall, 107
- overall for industrial fluids, 166
- overall in heat exchangers, 159
- radiation, effect of, 149
- turbulent flow in an annulus, 148
- turbulent flow in pipes correlation, 147

Henry's law and mass transfer between two fluids, 242
Higbie model for mass transfer, 227

Ideal gas equation, 6
Insulated pipes, critical radius, 112
Interferometer, optical for determining the heat transfer coefficient, 136

Kirchhoff's law, radiation, 173

Laminar flow in pipes, 55
Lewis number, 13
Lift force on a submerged body, 66

Manometers, 20
Mass transfer, convective:
- coefficient of, 225
- diffusivity D, 227
- Higbie's model, 227
- at a plane surface, 228
- single fluid phase, 224
- at a spherical surface, 231
- stagnant film model, 224
- stagnant film model, two fluids, 240

Mass transfer, natural convection:
- solution mining of potash, 237
- sphere, 235
- vertical plates, 235

Mass transfer by molecular diffusion (*see* Diffusion, molecular)
Mass transfer coefficient:
- for absorption of SO_2 in water, 247
- and diffusivity D, 227
- measurement in a wetted-wall column, 243
- overall, 241
- for packed columns, 247
- single fluid phase, 225

Minor losses in pipe flow:
- enlargement, sudden, 59
- conical expansion, 60
- contraction, sudden, 61
- pipe fittings, 61

Molecular weights, 258
Momentum integral, von Kármán, 88
Moody diagram for pipe friction, 51

Newton's law of cooling, 133, 134
Nikuradse, J., 95
Nozzle, 32
  head loss, overall, 36
Nusselt number, 142, 145

Orifice meter, 34

Packed column:
  mass transfer in, 247
  tower size to remove SO_2 from air, 250
Pitot tube, 31
Planck's equation for monochromatic emissive power of a black body, 174
Planck's radiation functions, 256
Prandtl, Ludwig, 71, 76, 99
Prandtl number, definition, 13
Pressure, fluid:
  definition, 18
  variation with elevation, 18
Pressure measurement:
  dynamic pressure, 30
  manometers for, 20, 30
  mercury barometer, 21
  stagnation pressure, 30

Radiant heat exchange:
  between black bodies, 189
  emissivity, definition, 173
  gray surface, definition, 190
  Stefan–Boltzmann equation, 173
Radiant heat transfer:
  absorptivity, definition, 171
  black-body, 172
  between a black body and black enclosure, 177
  electrical analog, 192
  emissive power, definition, 174
  geometric flux algebra, 185
  between gray bodies, 191
  Kirchhoff's law, 173
  in open systems, 177
  between two parallel black surfaces, 176
  Planck's equation for monochromatic emissive power of a black body, 174
  radiation intensity, 177
  radiation shielding, 198
  reflection from surfaces, 171
  reflectivity, definition, 171
  from reradiating surfaces, 197
  shape factor relationships, 181
  special reciprocity relation, 186
  spectral energy distribution, 175
  spectrum, 171
  transmissivity, definition, 171
  Wien's law, 174
Radiation intensity, 177
Radiation shielding, 198
Radiosity, definition, 192
Reflectivity, for radiation, 171
Reradiating surfaces, 197
Reynold's experiment, 22

Schmidt number, 13
Shape factor, radiation (*see* Configuration factor)
Solution mining of potash, 237
Spalding correlation (*see* Diffusion coefficient)
Stefan–Boltzmann equation, 173
Streamlines, 22
Systems of measurement:
  absolute metric system, 3
  British gravitational system, 3
  conversion of units, 4
  English absolute system, 3
  English engineering system, 2
  molar units, 4
  Système International d'Unités (SI), 3
  thermal units, 4

Thermal conductivity, 10
  values for industrial solids, 255
Thermal diffusivity, 11
Transmissivity, for radiation, 171
Transport processes:
  diffusion, Fick's equation of, 12, 205
  Fourier's equation of heat conduction, 10, 103
  Newton's equation of viscosity, 8
  rate of transport, 7

Venturi meter, 32
  head loss, overall, 38
View factor, radiation (*see* Configuration factor)
Viscosity:
  for air, 255
  kinematic, 9, 18
  Newton's equation of, 8
  units, 9
  for water, 254
von Kármán, momentum integral, 88

Water, properties of, 254
Wetted-well column, for measurement of mass transfer, 243
Wien's law, radiation, 174